Scaffolding the New Web

Standards and Standards Policy for the Digital Economy

Martin Libicki, James Schneider, David R. Frelinger, Anna Slomovic

Prepared for the
Office of Science and Technology Policy

Science and Technology Policy Institute
RAND

PREFACE

This work is RAND's response to a request made in FY 1999 by the Office of Science and Technology Policy (OSTP) to investigate the current standards development process to assess its adequacy and recommend public policies that may be warranted by the need to keep it healthy.

The resulting report aims to enhance the reader's sophistication about the standards process and its central issues. It is discursive but cannot claim to have discovered eternal principles. Why? The standards field is complex and nuanced and, like many an organic entity, looks more complicated as one draws closer. It is easier to explain what does not work (complex top-down specifications) than to guarantee that any alternative approach will, in fact, succeed—particularly when the market is molting as rapidly and repeatedly as it does. More so, the text is a snapshot in time. Judgments about the odds of this or that approach reflect the state of knowledge circa the summer of 1999 (and, of course, not the many twists and turns of the market since then). Thus, although this work can inform standards strategy in general, it cannot form the basis for a strategy for any *specific* standard.

The content should be of interest to members of the technology policy community and those curious about how information technology markets work. Readers are assumed to be generally knowledgeable about the industry's structure and products but not necessarily about information technology standards *per se*.

Originally created by Congress in 1991 as the Critical Technologies Institute and renamed in 1998, the Science and Technology Policy

Institute is a federally funded research and development center sponsored by the National Science Foundation and managed by RAND. The institute's mission is to help improve public policy by conducting objective, independent research and analysis on policy issues that involve science and technology. To this end, the institute

- Supports the Office of Science and Technology Policy and other Executive Branch agencies, offices, and councils
- Helps science and technology decisionmakers understand the likely consequences of their decisions and choose among alternative policies
- Helps improve understanding in both the public and private sectors of the ways in which science and technology can better serve national objectives.

Science and Technology Policy Institute research focuses on problems of science and technology policy that involve multiple agencies. In carrying out its mission, the institute consults broadly with representatives from private industry, institutions of higher education, and other nonprofit institutions.

Inquiries regarding the Science and Technology Policy Institute may be directed to the addresses below.

Bruce Don
Director
Science and Technology Policy Institute

Science and Technology Policy Institute
RAND
1200 South Hayes Street
Arlington, VA 22202-5050
Phone: (202) 296-5000
Web: http://www.rand.org/centers/stpi
Email: stip@rand.org

CONTENTS

FIGURES AND TABLES

Figure

Table

SUMMARY

With every passing month, the digital economy grows stronger and more attractive. Much, perhaps, most of this economy rests upon the Internet and its World Wide Web. They, in turn, rest upon information technology standards. Today's standards appear good enough to see the digital economy through the next few years. But it is unclear how much longer the momentum of such commerce can be sustained absent new standards. Are today's standards processes adequate? Where are they taking the industry (and where is the industry taking them)? Is government intervention required to address systemic failures in their development?

To answer these questions, a RAND Science and Technology Policy research team undertook five case studies covering

1. existing Web standards
2. the extensible markup language, XML
3. digital library standards
4. issues related to payments, property, and privacy
5. evolving electronic commerce value chains.

A White House–sponsored meeting of standards practitioners also generated material helpful in developing an overall assessment. All this material was used to inform the body of the report.

Information technology standards are a means by which two or more products (or systems) can function together. Some standards permit peers to interoperate or to exchange data in ways that are mutually comprehensible. Others permit one thing (e.g., a software application) to work atop another (e.g., an operating system). Information technology has seen a long march away from proprietary conventions (e.g., how the alphabet is translated into bit strings) toward open conventions that have become standards. The Internet and the World Wide Web exemplify openness; their standards are public and largely vendor-neutral. Yet as more products follow standards, innovative products are, almost by definition, unstandardized (in communications, standards often precede product introduction: One phone is useless by itself). So, the conflict between different ways of doing things starts anew. Standards failures tend to have one of three consequences:

1. New activities are stillborn.
2. New activities emerge, but with little interoperability among domains (each with its own conventions).
3. Proprietary standards enable an active but biased marketplace, reducing competition and hobbling innovation.

So far, the process by which standards are written and stamped remains basically healthy. True, the formal standards development organizations that were overtaken by the Internet Engineering Task Force (IETF) in the early 1990s remain on the periphery of the process; the IETF itself has become congested by its own popularity. But consortia (e.g., the World Wide Web Consortium) and forums (e.g., the Wireless Access Protocol Forum) appear to have picked up the slack. The rise of open-source software (e.g., Linux, Apache, Mozilla) has been another force for vendor-neutral standardization.

Yet, the case studies suggest that the success of standards in the marketplace depends on the play of larger forces. HTML and, to a lesser extent, Java succeeded because they were straightforward and unique ways of doing interesting things. But today's Web standards developments are wrapped up in the contests between corporations waging wars over browsers and other Web on-ramps, each trying to do an end-run around each other's proprietary advantages. The standards that would govern digital libraries, intellectual property rights, payments, and privacy are buffeted by the varied interests of affected groups—authors, librarians, rights holders, consumers, banks, merchants, privacy activists, and governments. Although

XML has quickly achieved wide acceptance, it is only the grammar through which Web content can be described. Many groups now vie to establish the words (i.e., the tag sets) everyone else will use: The result so far is a high head of froth and thin beer beneath.

The battle over tag sets reflects the broader problem of describing the messy real world to the sheltered naïfs that our computers still are. There is no obvious way to achieve semantic standardization. Creating one master tag set is optimal but a long shot. Creating tag sets specialized for various communities may be only somewhat more likely but complicates communicating *across* domains (each of which then also needs its own software). Translators would obviate the need for standards, but reliable translation exceeds what today's technology can provide. Ontologies into which everyone's terms can be mapped might improve translation, but how will a standardized ontology come about? Perhaps the best outcome is that some terms are globally standardized; some are locally standardized; and the rest are anyone's guess. There is, incidentally, little cry for the U.S. government to dictate what tags to use.

Does government, in fact, have much of a role to play? Standards for describing and measuring content (e.g., movie ratings, cyber-security performance) may substitute for some regulation. But less may be more: Many standards developers already believe that the government's overly liberal granting of patents on software (and business processes) frustrates the development of standards. Researchers might be allowed to use a fraction of their government research and development funding to work on standards. Perhaps the best help the government can offer is to have the National Institute for Standards and Technology (NIST—specifically, its Information Technology Laboratory) intensify its traditional functions: developing metrologies; broadening the technology base; and constructing, on neutral ground, terrain maps of the various electronic-commerce standards and standards contenders.

ACKNOWLEDGMENTS

This work was greatly aided by OSTP's sponsorship of a focus group on information technology standardization. This July 20, 1999, meeting assembled a distinguished group of standards developers: Tim Berners-Lee (World Wide Web Consortium), Scott Bradner (Harvard and IETF), Carl Cargill (author of *Information Technology Standards*), Robert Kahn (Corporation for National Research Initiatives), Shirley Hurwitz (NIST, Advanced Technology Program), Clifford Lynch (Coalition for Networked Information), William Mehuron (NIST, Information Technology Laboratory), Eric Miller (On-line Computer Library Center), Mary Mitchell (U.S. General Services Administration), Jim Ostell (National Institutes of Health), Michael Spring (University of Pittsburgh), Martin Tenenbaum (CommerceOne), Patrick Vittet-Phillipe (European Union), Laura Walker (OASIS), Daniel Weitzner (World Wide Web Consortium), Randy Whiting (CommerceNet), and employees of OSTP and RAND. Because participation at the focus group was on a nondisclosure basis, statements cited from the meeting are attributed to one or another anonymous "focus group member."

Special thanks go to Brian Kahin, formerly of the Office of Science and Technology Policy, for his sponsorship, advice, and support; Caroline Wagner and Bruce Don of RAND for their guidance; Walter Baer, Robert Anderson, and Willis Ware of RAND for their review and comments; and Allison Yezril, formerly of RAND, for her contributions to Chapter Five and her assistance in cover design.

GLOSSARY

Ada	A computer language mandated for use within the DoD and used widely in aerospace applications but rarely elsewhere
AES	Advanced Encryption Standard (a potential successor to the DES, but with 128-bit keys)
ANSI	American National Standards Institute (the supergroup under whom most formal American SDO's work)
AOL	America Online
API	Application portability interface (a piece of software, usually embedded in an operating system, that translates software code into a request for service)
ASCII	American Standard Code for Information Interchange (a 256-character code for converting letters, digits, and punctuation marks into eight-bit numbers)
BID	Business Interface Definition
BOV	Business operational view
bps	Bits per second
BSI	Business system interoperation
C++	A computer language in which most commercial software these days is written

CALS	Computer-Aided Logistics Support (a program to standardize the expression of all DoD technical manuals and related product descriptions)
CDMA	Code-division multiple access (a way of using a communication channel by assigning each call a quasi-random set of frequencies to use)
CD-ROM	Compact disk read-only memory
CEN	Committee for European Normalization (an SDO)
CEO	Chief executive officer
CERN	European Council for Nuclear Research (a high-energy physics laboratory in Geneva, Switzerland)
CIO	Chief information officer
COM	Common Object Model
CORBA	Common Object Request Broker Architecture (a suite of standards that allows applications to call objects that reside on networked servers)
DARPA	Defense Advanced Research Projects Agency (a DoD agency responsible for long-range research and development; it invented the Internet 30 years ago)
DES	Data Encryption Standard (an encryption standard that is symmetric in the sense that the same key is used for encoding and decoding)
DESIRE	Distributed European System Interoperability for Reinsurance (a consortium)
DIVX	Digital video express (a form of DVD)
DNS	Domain Naming Service (the mechanism by which alphanumeric Internet addresses are converted into unique numeric form)
DoD	U.S. Department of Defense
DOI	Digital object identifier (a proposed method of labeling intellectual material by content, and thus only once, rather than by location, as a URL does)
DSSSL	Document Style Semantics and Specification Language

DTD	Document type definition (a place in a document where the structure of tag sets is defined)
DVD	Digital video (or versatile) disk (a CD-ROM that, as currently implemented, holds 4.7 gigabytes of information)
ECMA	European Computer Manufacturers Association (an SDO)
E-commerce	Electronic commerce
EDI	Electronic data interchange (a generic name for any digital E-commerce data)
EDIFACT	EDI for Administration, Commerce, and Transport (an international EDI standard under United Nations auspices)
EDML	Electronic Data Markup Language
FSV	Functional service view
FTC	Federal Trade Commission (a U.S. government agency with some oversight over antitrust matters)
FTP	File Transfer Protocol (an Internet protocol)
GIF	Graphics Interchange Format (the format in which almost all Web imagery is encoded)
GILS	Global Information Locator Service (né the Government Information Locator Service)
GSM	Groupe Spéciale Mobile (a European-developed standard for digital cellular telephony)
HL7	Hospital Layer 7 (a standard used to exchange admission and billing information among hospitals and billing agencies)
HTML	Hypertext Markup Language (the language in which Web pages are written)
HTTP	Hypertext Transfer Protocol (the Internet protocol for moving Web pages between server and client)
IETF	Internet Engineering Task Force (a quasi-formal group under whose auspices Internet protocols are written)

INDECS	Interoperability of Data in E-Commerce Systems (a proposed standard way to tag information with appropriate intellectual property rights markings)
IPO	Initial placement offering (of stock)
ISDN	Integrated Services Digital Network (a suite of standards used to support digital telephony)
ISO	International Organization for Standards (an international SDO, under which sit national standards umbrella organizations, such as ANSI)
ITL	Information Technology Laboratory (that part of NIST responsible for information technology standards)
ITU	International Telecommunications Union (a UN-sponsored entity responsible for global telephone and television standards)
Java	A computer language used to write small applications (applets) that are uploaded onto client Web pages and run within them
JTC	Joint Technical Committee (an ISO working group)
JVM	Java Virtual Machine (a piece of software that runs Java applets)
MARC	Machine-Readable Cataloging (a standard way to express the contents of books for interlibrary loan purposes)
MIME	Multipurpose Internet Mail Extension (a way of converting 8-byte files, such as attachments, into files that can be carried within the body of an Internet E-mail)
MPEG	Motion Picture Experts Group (a compression standard for motion pictures and, incidentally, music)
NASA	National Aeronautics and Space Administration
NCSA	National Center for Supercomputer Applications (a laboratory associated with the University of Illinois, Champaign-Urbana)

NIH	National Institutes of Health (a U.S. government agency)
NISO	National Information Standards Organization (a U.S. SDO for librarians)
NIST	National Institute of Standards and Technology (a U.S. government agency)
OASIS	Organization for the Advancement of Structured Information Standards (a consortium that is attempting to build a repository of E-commerce tag sets)
OCLC	On-Line Computer Library Center (the central focus for interlibrary loans in the United States)
OO-edi	Object-Oriented EDI
OSI	Open Systems Interconnection (a suite of data communication standards that competes with TCP/IP)
OSTP	Office of Science and Technology Policy (the office of the U.S. president's science advisor)
P3P	Platform for Privacy Preferences Project (a proposed W3C standard that governs how servers and clients negotiate the terms under which personal information can be used)
PAS	Publicly available specification (an ISO device used to permit a standard written somewhere else to be crowned as an ISO standard)
PC	Personal computer
PDF	Portable Document Format
PGP	Pretty Good Privacy (a widely used but not formally standardized method used to encrypt E-mail)
PICS	Platform for Internet Content Selection (a proposed W3C standard language for describing content ratings, etc.)
PKE	Public-key encryption (a way of passing encrypted messages that permits encryption keys to be exchanged in the clear without risking interception of decryption keys)

RDF	Resource Description Framework (a proposed standard that would specify how tag definitions are made)
RFC	Request for comment (a series of documents that are either Internet standards or significant discussions of the Internet)
RIAA	Recording Industry Association of America (the trade group of the music industry)
SDLIP	Simple Digital Library Interoperability Protocol (a proposed standard that would govern electronic exchanges with libraries)
SDMI	Secure Digital Music Initiative (a proposed specification that would permit digital music to be sold in ways that would inhibit unlimited copying)
SDO	Standards development organization (usually refers to formally established standards groups)
SET	Secure Electronic Transactions (a standard, secure method of using credit cards over the Web)
SGML	Standard Generalized Markup Language (a way to mark up compound documents)
SMTP	Simple Mail Transfer Protocol (the Internet's E-mail protocol)
SNMP	Simple Network Management Protocol (the Internet's network management protocol)
SQL	Structured Query Language (a standard language used to formulate queries posed to databases)
SSL	Secure Socket Layer (a standard way to transfer secure information, such as payment data, over the Web)
STEP	Standard for the Exchange of Product Data (a standard used to format CAD/CAM files and express other sorts of manufacturing data)
TCP/IP	Transmission Control Protocol/Internet Protocol (the key transport and addressing protocol for the Internet)

TDMA	Time-division multiple access (a way of using a communication channel by assigning each call a sequence of time slices)
URL	Uniform resource locator (a unique address for Web content)
URN	Universal Resource Name
VAN	Value-Added Network
W3C	World Wide Web Consortium (a consortium assembled to create standards for the Web)
WAIS	Wide-Area Information Server (a method for indexing large amounts of document by included word)
X12	The ANSI committee that created a standard of the same name that specifies how business data are formatted for EDI.
XHTML	Extensible Hypertext Markup Language (an XML version of HTML 4.0)
XLL	Extensible Link Language (a proposed language for expressing links within XML documents)
XML	Extensible Markup Language (a now-common syntax used to mark up text for subsequent computer processing)
XML-QL	XML query language
XSL	Extensible Style Language (a proposed method for specifying style sheets that convert marked-up text into displayed text)
Z39.50	An NISO standard for queries of library catalogs

Chapter One

INTRODUCTION

> Openness is an underlying technical and philosophical tenet of the expansion of electronic commerce. The widespread adoption of the Internet as a platform for business is due to its non-proprietary standards and open nature as well as to the huge industry that has evolved to support it. The economic power that stems from joining a large network will help to ensure that new standards will remain open. More importantly, openness has emerged as a strategy, with many of the most successful e-commerce ventures granting business partners and consumers unparalleled access to their inner workings, databases, and personnel. This has led to a shift in the role of consumers, who are increasingly implicated as partners in product design and creation. An expectation of openness is building on the part of consumers [and] citizens, which will cause transformations, for better (e.g., increased transparency, competition) or for worse (e.g., potential invasion of privacy), in the economy and society.
>
> —*Organisation for Economic Cooperation and Development, 1999*

The digital economy sits at the uneasy juncture that separates the idealism of its youth from the moneymaking of its maturity. As a whole, it is *terra incognita*: Everything is new; the landscape is sure to change even as it is brought under the plow; and new standards are the throughways by which the favored few will reach farthest into new territory—or are they?

Perhaps new standards are not essential: The Christmas 1998 shopping season proved that the central question for electronic commerce (E-commerce) had shifted from "whether" to "how much, how soon" (and the 1999 season was more than twice as busy). But proponents of more advanced services, such as shopping (ro)bots, effortless E-currency, or search engines with more intelligence would argue that the "netizen" ten years hence will not be able to understand how people got along in 1999 with such primitive offerings.

If the digital economy requires new standards, the process by which they are formulated and disseminated becomes central to its prospects. Will it be well-served by today's standards processes—that is, will standards arise that are both well-conceived and timely?

This report seeks to shed some light on this question by successively discussing the place of standards (Chapter Two), lessons from five case studies (Chapter Three and Appendixes A through E), the emerging challenge of common semantics (Chapter Four), standards development institutions (Chapter Five), and public policy (Chapter Six). Chapter Seven presents conclusions, and Appendix F discusses the meaning of the term *standard*.

Chapter Two

THE PLACE OF STANDARDS

Five years ago, Wall Street, Silicon Valley, and Hollywood hoisted competing visions of the information superhighway. Many were backed by billions of dollars, whether from bonds, venture capitalists, or ticket sales. "Set-top boxes" were a popular focus.

The Internet, by contrast, had no such backers and modest governance. But it did have standards. And that was enough to prevail.

Conceived in the 1960s, the Internet was realized in the 1970s and early 1980s with the development and refinement of protocols for message transport (Transmission Control Protocol/Internet Protocol [TCP/IP]), file transfer (File Transfer Protocol [FTP]), E-mail (Simple Mail Transfer Protocol [SMTP]), and the ability to log onto remote systems (telnet). Such standards, coupled with a spare structure for addressing (Domain Name Service [DNS]), routing, and technology insertion (the Internet Engineering Task Force [IETF]), supplied the rules by which new networks could link themselves to the Internet and thereby exchange information with users on old networks and with each other.

It took standards from outside the IETF, however, to propel the Internet into today's prominence. The development, circa 1990, of the Hypertext Markup Language (HTML) and Hypertext Transfer Protocol (HTTP) provided a foundation for creating and transferring structurally complex documents across the Internet. Once graphical browsers appeared in 1992–1993 to take advantage of these standards, the Internet became visually exciting. The existence of display tools elicited content; with content came the demand for Internet membership and yet more tools.

The Internet and the World Wide Web, as it brought together disparate threads of information technology, also affected standards. Those compatible with the Web—such as Adobe's Portable Document Format (PDF), Compuserve's Graphics Interchange Format (GIF) for images, Motion Pictures Expert Group (MPEG) music compression, and Pretty Good Privacy (PGP) encryption—did well. Those left behind by the Web—such as computer graphics metafile, the American National Standards Institute's (ANSI's) X12 for electronic data interchange (EDI) for business, the Ada programming language, and Microsoft's rich text format—did not.

To get from the *present* to the *future* relationship of standards to the digital economy, it first helps to ask what standards do.[1]

WHAT MAKES A STANDARD STANDARD?

Computers, swift but stupid, are poor at inferring what something—a program, a user, another computer, a network—means, as opposed to the ones and zeroes actually used to convey data. Information and information-transfer mechanisms must therefore be composed in precise and mutually understood terms. If a convention for doing so is sufficiently common, it can be called a standard. An imprimatur of such a convention from one or another standards development organization (SDO) is not necessary but does help. Formal standards descriptions tend to be rigorous and clearly spelled out (particularly when contrasted to proprietary conventions).

A convention may be judged by its technical merits: Does it solve a problem? Does it do so elegantly? Is its solution clear? Is it easy to implement? Is it powerful enough to permit users to do what they want to do? Can its correct use be easily tested? A standard may also be judged by the fairness of the process in which it was developed: SDOs are also pickier about due process, which makes their products formally reviewed and, some believe, more fair.

---

[1]Considerable work on standards theory was undertaken in the late 1980s and early 1990s. See David and Greenstein (1991), Spring (1991), and *Information Infrastructure and Policy*'s special issue on interoperability (1995). For a broader perspective on standards and the digital economy, see Shapiro and Varian (1999).

Nevertheless, the true test of a standard is that it be widely used. The wider the use, the lower the cost of interoperation between two random users, the more people and processes that can interact with each other, the less the need for translation (and the inevitable loss of meaning) to exchange information, and the greater economies of scale in producing support services, tools, and training. Once a convention becomes a true standard, alternatives tend to lose support—leaving some users worse off (e.g., even in a world where C++ dominates, other computer languages, such as Ada, have their unique strengths). Furthermore, the best conventions do not necessarily graduate to standards: Those that win early acceptance or are merely crowned by the expectation of success may attract the next wave of users who want to interoperate with as many prior users as possible (i.e., the "network effect") or who at least do not wish to be stranded down the road. The more users, the greater the expectations of further success. And so on.

There are essentially two approaches to standardization. *Minimalists* value simplicity and rapid uptake by the user community. Their standards tend to be expressed as primitives from which subsequent elaboration takes place *after* acceptance occurs. Theirs is an inside-out world. *Structuralists* value comprehensiveness and precision in the fear that rough-and-ready standards will, at best, grow like weeds, making well-kept ontological gardens that much harder to maintain. They would model the world so comprehensively that no human activity, extant or imagined, would fall outside their construct.[2] Such activities are then mapped into successively finer categories of relationships, which are then enumerated and labeled. Theirs is an outside-in world.

Open Systems Interconnection (OSI) is clearly structuralist: It grew from a reference model that partitioned all data communications into seven layers, from the physical exchange of bits to the organization of data via applications (e.g., E-mail). The OSI reference model is universally acknowledged and rarely followed as such. The Common Object Request Broker Architecture (CORBA) is another similarly ambitious (albeit less structuralist) set of standards for

[2]A typical structuralist approach is the Universal Modeling Language, a spin-off from the Ada computer language community.

object-oriented middleware that would rope in an enterprise's legacy base of software and hardware.

The Internet, by contrast, was created by minimalists who forswore grand conceptions to focus on a few good protocols (e.g., TCP/IP) that would permit the services they wanted. It handily beat OSI at its own game. Although the Web's creators may have sought a comprehensive structure to the universe of documents (see Berners-Lee, Connolly, and Swick, 1999), HTML rose to prominence as a set of well-chosen primitives rather than the expression of any such structure.

Successful standards, correspondingly, tend to start small, not large. The entire C reference manual fills no more than 40 pages of broadly spaced print. (Kernighan and Ritchie, 1978, pp. 179–219.) The first version of HTML could be learned, in its entirety, in an hour. TCP/IP, and the Structured Query Language's (SQL's) rules could be stated very succinctly. By contrast, very complex standards—such as OSI, Ada, Integrated Services Digital Network (ISDN), the Standard for the Exchange of Product model data (STEP)—were born large and complex; the first two have largely failed, and the second two are struggling. Of course, neither C (which grew to C++) nor HTML *stayed* simple, but both caught on before they evolved toward greater complexity.

Standards are shaped by conflicts within and between communities of computer engineers and corporate representatives. Engineers prefer standards be elegant and functional; aesthetic differences often lead to fierce standards fights arcane to outsiders. Corporations cooperate to use standards for new markets, but vie over their details to widen or narrow access to existing ones (depending on who is on top). Engineers bickered over the SONET fiber-optic trunk-line standard until corporate executives commanded them to get on with finding a standard so that firms could interconnect. Standards also permit many technical features of a product to be described in shorthand, leaving companies to play up its unique features. Engineers alone battled over whether Open Step or Motif would become the standard graphical user interface for X-Window/UNIX systems—until corporations realized that user interfaces were a useful way to differentiate workstations. Then distinctions were emphasized in public.

Browsers appear to play a key role in validating Web standards—in the sense that an innovation not supported by a browser is in trouble. For the nonce, browsers remain the door to the Web and hence the digital economy.[3] Not for nothing has Microsoft's assault on the *browser* market been front and center in its antitrust case—even though this one product contributes but a small fraction of its business and makes little money on its own. But how much influence do browser companies wield in the overall process? Java caught the imagination of developers before it showed up in browsers. Yet, if Java had not shown up in one soon enough, it would have died. Conversely, once Java had enough momentum, any browser that did not support it would have hurt itself. Clearly any *new* version of Java or HTML (or its putative successor, the Extensible Markup Language [XML]) *not* supported by a popular browser has a poor chance of success. Assisting the browser is a vast array of plug-ins with conversion, display, and manipulation capabilities (some of which, such as Adobe's Acrobat software for reading PDF files, antedate the Web).[4]

THE POTENTIAL IMPORTANCE OF STANDARDS

Standards failures leave up to three problems in their wake: (1) New activities are stillborn; (2) new activities emerge but with little interoperability among domains that follow their unique standards; or (3) proprietary standards enable a thriving but biased marketplace, thereby reducing competition and ultimately retarding innovation.

Good standards clarify investment decisions. Since everyone uses TCP/IP for packaging Internet content, engineers understand what they have to engineer their networks to do. Those who generate content or support services, in turn, know what they have to break down their information streams into. Network providers need not worry so much about what kind of content they are carrying, and

---

[3]But not the only door. Popular techniques for real-time audio and video streaming do not work through browsers; neither do downloads to palmtops. Instant messaging has also been viewed as a new portal into the Web; see Paul Hagan of Forrester Research, as quoted in Ricciuti (1999).

[4]A plug-in is a piece of software that a browser loads to perform a specific function. Netscape's Web site (http://www.netscape.com/plug-ins/index.html) listed 176 external plug-ins as of March 16, 1999.

content providers do not need to worry about what networks their content flows over. Consider Figure 1. Multiple information applications, services, and formats are listed on the top, and multiple network technologies are listed on the bottom, but only one transport-cum-addressing service occupies the middle. The latter is clearly TCP/IP (and/or descendants). TCP/IP's achievement was to simplify an otherwise exceedingly complex three-dimensional mix-and-match problem into a tractable, two-dimensional mix-and-match problem.

TCP/IP also illustrates how architecture emerges from standards. Both the packet switching of TCP/IP and the circuit switching of telephony can route messages, but they lead to different kinds of networks. Circuit switching, with its limited and controlled data-stream handoffs and its parceling of bandwidth in discrete units (e.g., of 64,000 bps lines), facilitates per-use billing and system management but frustrates the carriage of bursty data flows and high-bandwidth multimedia (which require bundling and synchronizing multiple lines). The multiple and globally unpredictable handoffs of discrete TCP/IP packets complicate per-use billing and system management, but TCP/IP is tailor-made for higher bandwidth. Telephony concentrates intelligence at the switch; packet switching concentrates intelligence at the terminal. (See Isenberg, 1997.) More generally, packetization obviates worry about which bit of content (e.g., voice, video, and data) uses which internal channel (e.g., which time-slice, or nth bit of 16). Going farther, markup languages (such as XML) permit structured content to be expressed without worry over what position or how many bits a particular datum occupies.

Cellular telephony illustrates the power of standards even over technology. In 1983, AT&T's roll-out of an analog standard (Advanced Mobile Phone System [AMPS]) kick-started cellular telephony in the United States. But Europe, with its multiple cellular standards, reaped confusion. So vexed, and anticipating a second generation of cellular systems based on digital technology, European countries agreed to develop a common system (Glenn et al., 1999) and, in 1991, deployed Groupe Spéciale Mobile (GSM), a time-division multiple-access (TDMA) standard. Meanwhile, in the United States, the absence of a mandated TDMA standard for cellular phones allowed Qualcomm, a start-up, to introduce, in 1990, another convention for

The Open Data Network

RANDMR1215-1

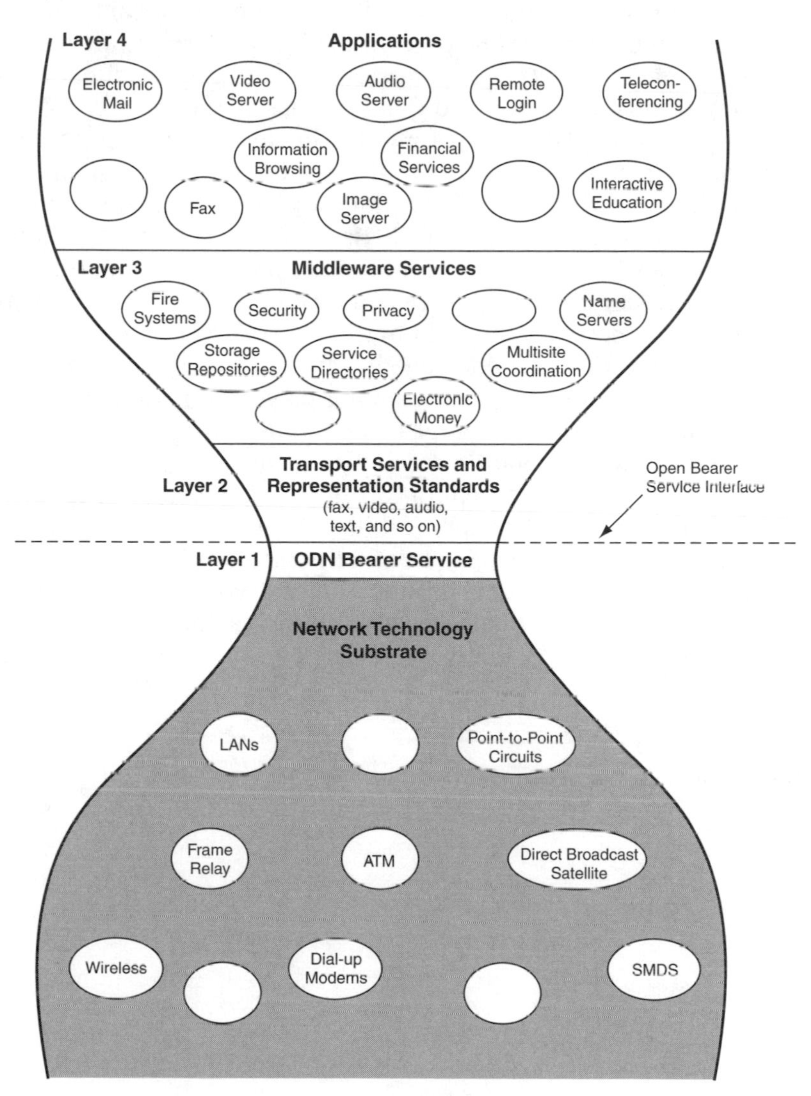

SOURCE: Reprinted with permission from National Research Council (1994), p. 53. Copyright 1994 by the National Academy of Sciences. Courtesy of the National Academy Press, Washington, D.C.

Figure 1—A Four-Layer Model for the Open Data Network

digital telephony, code-division multiple access (CDMA). It offered greater security and capacity (partially by using statistical multiplexing to exploit the fact that 60 percent of all voice circuits are silent at any one time). TDMA vendors responded with new frequency allocation methods that promised great increases in capacity.[5] Whose philosophy won? GSM was a great fillip to cellular telephony in Europe, permitting a level of continent-wide roaming long unavailable in the United States (where older analog systems remained in use). Furthermore, because GSM is a *global* standard, whereas Qualcomm's CDMA was but one of many *national* standards, GSM phone users could roam overseas as well. As of 1998, GSM had claimed 64 percent of the world market—well over 90 percent outside North America. Cellular telephony stands out as a high-technology market unique because its major players, Ericsson and Nokia, are European. Europe also appears ahead in putting Web access on cell phones. The International Telecommunications Union (ITU) is now working on a third generation of mobile systems capable of raising bandwidth up to two megabits per second. As 1999 ended, a compromise between Qualcomm and Ericsson left a CDMA proposal as the likely choice—even though the last standard was TDMA. There is little evidence that Europe's ability to achieve market dominance based on TDMA is any bar to their being able to do so again with CDMA.

---

[5]See, for instance, Therrien (1992).

Chapter Three

LESSONS FROM FIVE CASE STUDIES

Strong examples, such as the two just discussed, come from the world of communications infrastructures, where horizontal interconnection is a sine qua non of the business, and poor bets can cost companies billions of dollars. Physical infrastructure does not seem to be an E-commerce barrier, but the same may not necessarily hold for the semantic infrastructure (the encapsulating of business concepts into terms recognized by computers).

Approaching the issue of Web standards required doing four case studies on the present and one more on the future. The first examines two key components of the Web page, HTML and Java. The second focuses on XML and how markup may be used to bring order not only to the Web but also to E-commerce. The third discusses the raft of standards proposed to organize knowledge. The fourth deals with payments, privacy, and the protection of intellectual property. The last looks at the future of standards as a function of the still-evolving value chains of E-commerce.

These case studies both reinforce what decades of prior standards have already proven and acknowledge new requirements for external interoperability and a reasonable intellectual property regime. Five lessons merit attention.

STANDARDS FOSTER OPENNESS

The story of Web, E-commerce, and knowledge organization standards proves again that standards, regardless of how earnestly people try to manipulate them, are a force for openness. Where standards are absent, or ill-suited for their task, markets are closed or constricted, and raw market power prevails. Complaints about bias in the standards process are essentially secondary unless bias prevents the standards process from functioning at all. True, many standards battles (Netscape versus Microsoft on HTML, Sun versus Microsoft on Java, America Online [AOL] versus Microsoft on instant messaging) stem from disputes between Microsoft and an opposing coalition. And neither monopoly control nor hostile bifurcation is necessarily desirable. Nevertheless, it is difficult to find a de jure standard, howsoever skewed in development, that enshrined a market leader as well as closed or de facto standards have done.

BUT STANDARDS HAVE TO SOLVE PROBLEMS, BOTH TECHNICAL AND SOCIAL, TO SUCCEED

The life span of standards can often be predicted by gauging what and whose problems they solve. Thus, Secure Electronic Transactions (SET—a payment mechanism) has lagged because consumers have not been convinced they needed its authentication services; micropayments have lagged because consumers have not been persuaded to pay for information or, at any rate, not in dribs and drabs. HTML permitted users to look at documents as they access them—something FTP alone did not provide. Java was a standard in search of a market, and once the expression of animated GIFs was standardized, it had less to offer the Web. This rule will doubtlessly apply to standards for software agents: Is this something customers need? Standards for knowledge organization illustrate the *whose* aspect: Can the needs of librarians persuade authors to categorize their works? Is the interlibrary loan model relevant to digital material? Can lawyers persuade publishers to identify the property rights inherent in a work?

THE INTERNET AND WORLD WIDE WEB HAVE SHIFTED THE FOCUS OF INTEROPERABILITY

In the 1980s and early 1990s, many firms that automated their departments separately found themselves with a large headache when building an enterprise system from them. The Standard Generalized Markup Language (SGML—a way to mark up compound documents) and CORBA (a way to build applications from components held by a network) were touted as middleware glue. Today the emphasis is shifting to linking with external customers and suppliers. Middleware has proven too heavy for external systems; ANSI X12 is suitable only for repeat business along well-established lines. Hence the popularity of XML, which lightened SGML and does not assume the existence of middleware or even that external users will employ common practices and models. XML has started to replace CORBA as a syntactic layer for standards ranging from CommerceNet's ecoSystem, Hospital Layer 7 (HL7), and three standards from Case Study 3 (Simple Digital Library Interoperability Protocol [SDLIP], the Dublin Core, and PubMed).

LIGHT STANDARDS CONTINUE TO DO BETTER

The simplicity of HTML, Javascript, and the Secure Socket Layer (SSL) has prompted their uptake on the Web. XML, by simplifying SGML, has given markup a great lift. The Dublin Core looks light enough to succeed. By contrast, SET and many of the proposed knowledge organization standards appear too heavy for takeoff, and the complex structural models being built for RDF (resource description framework) or used to bulwark future object identifier models do not feed optimism about either.

BUT THE ENCAPSULATION OF THE REAL WORLD INTO STANDARD SEMANTICS IS LIKELY TO BE DIFFICULT

With the enthusiastic adoption of the metalanguage, XML, issues of syntax, the easy work of standardization, appear settled. The gradient ahead to semantic standards is far steeper, with no obvious trail upward. This is because semantic standards are an abstraction of a complex universe. Backers of EDI/X12, HL7, and perhaps the Dublin Core must hope the semantic structures and implicit business models of earlier standards may be converted into straight semantics. Otherwise, common *notations* overlaid upon dissimilar *notions* of how the world of discourse is constructed will lead to ambiguity: messy for humans and dangerous for machines.

Indeed, the search for semantic standards is becoming the touchstone for all upper-level standards efforts. How to reach that goal merits consideration in its own right.

Chapter Four

THE EMERGING CHALLENGE OF COMMON SEMANTICS

With XML has come a proliferation of consortia from every industry imaginable to populate structured material with standard terms (see Appendix B). By one estimate, a new industry consortium is founded every week, perhaps one in four of which can collect serious membership dues. Rising in concert are intermediary groups to provide a consistent dictionary in cyberspace, in which each consortium's words are registered and catalogued.

Having come so far with a syntactic standard, XML, will E-commerce and knowledge organization stall out in semantic confusion? With at least one human taking part in every transaction, business-to-consumer commerce should not be greatly affected (poor prospects for shopping bots may not bother site owners that profit from strong brand loyalty). But standardization matters greatly for business-to-business commerce, with its repeat purchases, steady cost pressures, and potential savings from tying purchasing to automated production and scheduling systems. This also holds for knowledge organization, with many subject areas supported by literally millions of documents.

How are semantic standards to come about? Five paths are suggested below.

LET THE MARKET DECIDE

At first, multiple standards consortia create competing vocabularies, some better than others. Confusion reigns. Many small clusters latch onto one or another standard; others follow the dictates of their primary client. Everyone else, paralyzed by the many choices, watches and waits. In time (perhaps only Internet time), momentum develops for a preferred tag set. Once this momentum is recognized, competing alternatives are discarded, and consolidation proceeds rapidly. Everyone ends up speaking the same language.

That is the happy version. It appropriates the advantages of natural selection in that the fittest survive, and no big brother, whether public or private, need intervene with a heavy and perhaps clumsy hand (alternatively, such intervention starts, but the fingers work so slowly that consolidation takes place before the grip is tightened).

But is the happy version likely? A chicken-and-egg cycle may yield paltry results: E-commerce remains a manual undertaking without universal standards with which to program computers, and the forces that would foster consolidated standards work without great urgency because the applications that need such standards are not imminent. E-commerce clusters may even form around a dominant buyer or vendor (e.g., for office supplies—although the Open Buying Initiative is headed by an Office Depot vice president), duplicating in cyberspace the kind of *keiretsu* that the Japanese invented for real space. Competing clusters that form at the national (or linguistic) level may, ironically, retard today's healthy progress toward a global economy. It is unclear whether such clusters would be precursors or barriers to eventual consolidation.

Granted, standards could consolidate too fast without adequate consideration of alternatives. But a greater threat arises because the ecology of standards is anything but natural. Absent standards, profits await institutions that can shepherd the bulk of transactions under their roofs for a small fee. Even a standard born of proprietary instincts may foster a monopoly over critical aspects of E-commerce. Worse may result if the winner is already a full or near monopolist in an ancillary field (e.g., office software, on-line Internet provision), so that one monopoly position reinforces another.

HAVE DIFFERENT COMMUNITIES EACH DECIDE

A variant of the Darwinian struggle is that each sector generates a common vocabulary for its own business based on its own standards work. Smaller, more homogenous groups may succeed where larger ones fail.

Would sector-specific standards suffice? The boundary between sectors has never been easy to delineate and is not always meaningful. Axles and tires are sold to automakers; tires and uniforms, to department stores; uniforms and medical services, to hospitals. Many large customers—not least, the federal government—would have to conduct E-commerce across a wide span of sectors. Because every standard rests on its own business model, which is reified in complex enterprise management software, multiple business models make business process systems unwieldy to create, operate, and maintain. The digital economy is redefining communities anyway. Who would have thought that the orderly business of bookselling and the chaotic business of auctioneering would have Internet business models with such common features? If the spirit of the Internet is universality, why settle for standards with the opposite effect?

ASSUME INTELLIGENT SOFTWARE WILL MEDIATE AMONG VARIOUS VOCABULARIES

Conceding diverse tag sets, another approach to E-commerce and knowledge organization would have sophisticated software mediate among them, much as people who speak different languages can be understood through translation.

If the standards problem were no more than a simple one-for-one substitution ("you say tomato . . .") this approach could work. But translation presumes a common cognitive model of the universe described by various words. Uniformities on the structuring of text make it possible (if not easy) to translate between documents produced by Word Perfect and by Microsoft Word. Greater variations in the structure of graphical files make similar translation between Harvard Graphics and Microsoft PowerPoint nearly impossible. A resident of Calgary may have an easier time referring to winter in a conversation with a Quebecer, despite the difference in language, than in doing so with a Houstonian whose climate model differs greatly.

As a point of departure, the current U.S. EDI standards (ANSI X12) speak both to common business processes (e.g., invoices) and industry-specific ones (e.g., for perishables, automobile parts, and hospital services).[1] Some business processes (e.g., invoices, again) are well-established but not in detail and not all. The simple concept of "offering price" may represent a constant in one model, a variable as a function of quantity in another, a variable as a function of timeliness in a third, and so on. An expiration date may have different meanings in food and photographic film (when discounting starts), in pharmaceuticals (when sales must end), and in software (when the sample ceases to work unless a key is purchased).

It is also unclear how tolerant business people will be for imperfect translation. Translation software is bound to be extremely sophisticated, and debugging it thoroughly may take years—and even then may not be entirely trusted.[2]

DEVELOP STANDARD ONTOLOGIES INTO WHICH STANDARD TERMS ARE MAPPED

Can disparate vocabularies be resolved through an ontological framework upon which each one would rest and to which each would refer? Such work is going on now (e.g., Ontology.org). Yet, finding middle ground between too little work on the area (indicating little interest) or too much work (indicating irreconcilable products at the end) is hard. Further, will the practical types that now go to standards groups be of a mind to profit from the work of the academic types that used to go to standards meetings and are still attracted by the high cognitive efforts entailed in building ontologies?

The quest for a philosophically clean language dates back before Ludwig Wittgenstein mooted the possibility in his first masterwork, *Tractatus* and conceded defeat in his last, *Philosophical Investigations.* The Defense Advanced Research Projects Agency (DARPA) has been investing in ontological development through most of the 1990s. Researchers initially optimistic about translation came to

---

[1]But traditional EDI (as Appendix B notes) is expensive to set up; is costly to operate (especially if it requires joining a proprietary value-added network); and, as a result, was hardly universal, even at its peak.

[2]Tim Berners-Lee has argued that common semantics may be inferred, in part, through analytic engines that can comb the Web and see how terms are used. (See Berners-Lee, 1999, pp. 177–196.)

believe that translation was likely to be adequate only within specific domains (e.g., answering weather and travel-related inquiries)—which seem to grow narrower with every reconsideration.

CONCENTRATE ON THE KEY WORDS

A compromise is to concede that, at best, some semantic primitives will be widely understood; others will be understood within specific communities; and the rest will have to be negotiated based on commonly accessible references.

Here, the broad standards community would seek consensus on what these primitives should be and how they should be defined. If and as standards take hold, they can be expanded outward. The prospects of success may be gauged by the record of successful standards that started small and grew rather than those that were born complex.

But prospects are not guarantees. There still needs to be some forum through which agreement can be sought on *two* levels: what is to be standardized and how. It is also unproven that there is a core set of E-commerce words that is small enough to be tractable for standardization purposes, common enough among the variegated world of business models, and yet large enough to encompass most of what a minimally useful E-commerce transaction must contain.

CODA

One possible approach, which is to have the federal government drive a solution through dictate or buying power, is simply on no one's agenda. No one is asking for it, least of all those most active in the various standards processes; the government's track record of championing specific standards is, at best, uneven (e.g., continuing to back OSI as the world turned to TCP/IP); and, although the government is toward the front of the E-commerce parade, it is not *at* the front. Explorations conducted by networks of interested people scattered throughout the bureaucracy are a far cry from having a coherent policy and direction. Even European governments, historically more eager to take charge (and whose purchases account for a larger share of their region's gross domestic product) have been holding back, waiting for their private sector to work its way toward standard.

Yet, as a practical matter, the ability of industry to develop coherent semantic standards depends on the health of the standards process.

Chapter Five

STANDARDS DEVELOPMENT INSTITUTIONS

Entrepreneurs propose, and the standards bodies dispose, tidying up the chaotic effusions of this or that brainstorm so that mutual comprehension may reign—in theory. But is the ecology of standardization as healthy as it should be? Can standardizers still rise above the tumult of competition without ascending the sterile heights of irrelevant perfectionism? How well, in fact, *do* today's standards organizations—from United Nations (UN)–sponsored groups to ad hoc consortia—work?

The victory of the Internet over OSI in the early 1990s did lend a retrospective aura to the Internet's build-a-little, test-a-little standards processes compared to the International Organization for Standards' (ISO's) more formal habits. Some of the comparison may have been unfair: HTML and XML were based on SGML, a bona fide ISO standard, and Javascript found a home at the European Computer Manufacturers Association (ECMA), an SDO. But the common wisdom persists. Does it still hold?

THE IETF

When the tide shifted from OSI to the Internet, the attention of business shifted as well. Five to ten years ago, IETF standards meetings were dominated by academics and other computer scientists; these days, businesspeople are likely to make up the overwhelming majority of participants—even where the subject is libraries. In 1987, the IETF's semiannual meetings had only 100 attendees (up from 15 a year earlier) and for at least five years afterward hosted a community whose members knew each other. With little money at stake, participants largely represented themselves. Agreements, while never easy, benefited from rapid feedback and ready bench-level testing of concepts.

Today, most IETF participants represent large concerns (some with stratospheric market values). Technical details are no longer so technical. Two thousand people attend the semiannual meetings. The IETF itself has been moved under the aegis of the Internet Society, which has self-consciously made itself international in recognition of often-different perspectives overseas.

Predictably, the IETF slowed down. The growing crew of network designers, operators, vendors, and researchers collectively created a bottleneck, preventing the rapid movement of standards. In theory, IETF standards processes are expeditious: six plus months for the Internet community to comment on proposed standards before they become draft standards and four plus months more until actual promotion to a standard. The effective time span is now longer. Between 1993 and 1999, it took roughly 3 years for a proposed standard to become a draft standard, and 5 years for a proposed standard to become a standard.

As Table 1 indicates, the number of proposed standards has increased threefold every three years since the mid-1980s, and number of draft standards rose similarly until the mid-1990s before level-

Table 1

Internet Proposed, Draft, and Final Standards (by year)

	Standard	Draft Standard	Proposed Standard
–1980	1		8
1981–1983	12		2
1984–1986	11		2
1987–1989	16	4	10
1990–1992	7	10	34
1993–1995	9	40	107
1996–1998	6	31	263

NOTE: For the *latest* version of the standard.

ing off. Meanwhile, the actual total of standards is, if anything, off a bit from its 1980s pace. The IETF created a total of 40 standards between 1981 and 1989. Yet, only 22, roughly half as many, were passed between 1990 and 1998—despite the pace of technological breakthroughs and the birth of the Web.

The IETF currently has more than 100 working groups and is continuously forming new ones but with no corresponding increase in the number of standards. Despite intense debate for or against the various protocols[1] and IETF's motto of "rough consensus, running code," the requirement for rough consensus has lead to splinter groups and yet more delay.

Such slow responses have sent many participants looking for a better way at a time when quick decisions are needed to keep pace with the burgeoning field of E-commerce. In 1996, a working group on Simple Network Management Protocol, Version 2 (SNMPv2) disbanded after a heated disagreement involving criteria for security and administrative standards. The ripples, felt throughout the computer-programming community, kept companies from implementing new versions of SNMPv2. Three years later, private vendors could only hope that the IETF's reassembled group will be able to endorse the 1998 SNMPv3 proposed standard. (Duffy, 1998.)

The IETF has responded by emphasizing the openness of the Internet community, the ability of that community to comment freely on issues that directly affect it, and the role of debates in weeding out inferior technology and providing technically superior standards. Recently, the debate process has seen hints of governance: As the Internet has grown, people unacculturated by their predecessors have continued to put forward their opinions in mailing lists, even after decisively hostile review; starting in 1998, people have been dropped from such lists.

Overall, the IETF has evolved away from being the progenitor of standards to the body that brings concepts into consensus. HTML and HTTP, as noted, arose from *outside* the IETF.

---

[1]Phillip Gross, a former chairman of IETF, interview in MacAskill (1988).

ISO, ITU, AND ECMA

International standards organizations, having been overtaken by the IETF, are trying to hasten their standards processes. The ISO has adopted a Publicly Available Specification process (see Appendix A) through which standards blessed in another forum can be speeded through to final ISO imprimatur. It has yet to see wide usage.

After 1992, the ITU ended its rule that standards be approved only in Olympic years. In special cases, standards can travel from proposal to imprimatur in five months, and nine months for others is not unheard of—difficult to imagine unless the standard is fairly well cooked before it enters ITU's kitchen. Perhaps the most interesting battle shaping up concerns Internet telephony (indeed, the very name bespeaks the clash of two cultures). The IETF's Simple Internet Protocol Plus (SIPP) draft standard and the ITU's H.323 specification, while using a similar architectural model (the Internet's Real-Time Transport Protocol [RTP]), are quite dissimilar in their details.

In Europe, ECMA has evolved into a forum in which competitors to Microsoft can try to coronate a standard in ways they could not at home (see Appendix A). The Committee for European Normalization (CEN), another European SDO, is easing out of the standards business for E-commerce and is testing a new role: convening workshops to identify areas of informal agreement and best practices. Under CEN's umbrella, the European Commission has launched a project to promote a project called Electronic Commerce Open Marketplace for Industry, with workshops under way or in preparation in such areas as sanitary wares, hospital procurement, construction, and textiles (European Union [EU], 1999, p. 9). CEN will help operate the workshops and provide them neutral technological expertise.

THE WORLD WIDE WEB CONSORTIUM (W3C)

The W3C is the closest analog to the IETF in the realm of Web (as opposed to Net) standards and the dominant force in XML standards development. Founded in 1994 by Tim Berners-Lee, it has several hundred members (mostly corporations), who have to pay dues. However, as with the Internet, the primary influence is exercised within the various working groups, which create and publish technical specifications. When officially approved by the W3C process, these specifications are considered tantamount to official standards.

The W3C has taken at least one media hit because attendance at its annual meetings has flagged. (Garfinkel, 1998.) However, it is firmly in control of its realm—the development of syntactic standards. Such standards ride atop the more bit-oriented standards of the IETF, although the exact boundary between the two is undefined (e.g., who builds the next version of HTTP). Meanwhile, the job of building semantic standards to exploit the W3C's syntactic standards is the province of consortia, such as OASIS.

THE WIRELESS ACCESS PROTOCOL (WAP) FORUM

The ecology of standards is populated by start-up consortia of multiple sizes and various life spans. The 147-member[2] WAP Forum illustrates some typical features. It was founded in 1997 to foster the use of browsers for cell phones. The world's big-three cell-phone makers—Ericsson, Nokia, and Motorola—were founders; a fourth, Phone.com, is a start-up that actually wrote the standards. The forum's literature emphasizes that it is not a standards group (but it has a three-stage specification approval process), that it will in due course submit its recommendations to SDOs, and that it liaises with SDOs and non-SDOs (e.g., the IETF, the W3C) alike. U.S. companies constitute less than half of the forum, and its standards style reflects this. Standards are layered (as were OSI's), are middleweight in complexity, and reflect key architectural assumptions (e.g., that cell phones have keypads, receive information in "cards," but do not talk directly to Web servers). Palm Computing, whose Palm VII has a different architectural model, was a notable latecomer to the group.

OPEN-SOURCE SOFTWARE

The last few years has seen the rise of open-source software, notably Linux (a UNIX-like operating system with roughly a third of all Web server operating systems), Apache (Web server software that has just over half of its market), and Mozilla (the open incarnation of Netscape's browser). If popular, their presence may complicate the process by which dominant firms leverage monopoly control of key software to further close off upstream applications from competition.

---

[2]As of November 23, 1999; see the forum's Web site (http://www.wapforum.org).

Is open sourcing a standards process per se? The source code of Linux embodies the language as a standard.[3] Its kernel has many contributors whose proposals are rigorously and enthusiastically vetted by peers from around the world. In the end, however, one person (Linus Torvalds) decides what is included. Outside the kernel, the process is more diffuse. Usually, the person who intuits the need for this or that extension gets to decide its contents (but, again, in open forum). Social mechanisms limit the degree of forking (two groups with incompatible approaches to a problem). Even so, fights have taken place between KDE and Gnome over which becomes the preferred user interface within the Linux community.

Open-source software is also no ironclad guarantee against market power. A friendly user interface, reliable hooks to the rest of a user's system, and hand-holding still play large roles in selling software. Red Hat Software has the largest share of the U.S. commercial market (even if many copies of Linux are downloaded for free) and enough "mindshare" to charge premium prices as well as to launch a successful initial placement offering (IPO).

Open-source software has a distinct advantage in that it permits users to modify the operating system to their specific needs. This is of special relevance to the federal government (notably the Department of Defense [DoD]) in its search for greater information security: Open sourcing not only allows bugs to be fixed quickly but permits an institution to release to its users a standard version with tempting capabilities removed and all the controls set correctly.

A TYPOLOGY

Open-source software raises a larger question: To what extent can open transparency in the creation of a standard substitute for more legal definitions of *fair* and *democratic*? Table 2 is a two-by-two typology of standards institutions. It suggests that having a strong

---

[3]Eric Raymond, admittedly a partisan, has compiled three papers on Linux: "The Cathedral and the Bazaar" (Raymond, 1998a), "Homesteading the Noosphere" (Raymond, 1998b), and "Open Source Software: the Halloween Document" (Raymond, 1998c).

Table 2

A Typology of Standards Organizations

	Democratic	Strong Leader
By membership	ISO	W3C
Open to all	IETF	Linux open source

leader may be an advantage, but so, to a lesser extent, is a process open to all (although the W3C is a membership organization, it does publish its draft specifications for open comment).

Standards development is a very pragmatic process, and there appears to be little barrier to new forms arising as need dictates. Ten years ago, virtually everything of note was done through formally established standards development organizations. Three new forms now vie for contention—plus the evanescent small-group consortia growing prominent in the E-commerce arena.

It would seem that the production of standards is the metric by which standards groups should be measured. But in times past, standards groups, notably those of a structuralist bent, also built intellectual foundations for subsequent standards, often generated by others. The OSI reference model is an example. Ontologies may provide a future one. Their value may be a question of timing. Information technology, like many human endeavors, undergoes cycles of efflorescence (from the early 1940s to the early 1960s and from the early 1980s onward) followed by consolidation (from the early 1960s to the early 1980s). A new wave of consolidation is inevitable. When its time comes, the existence of reference models may help promote standardization. But if standards development organizations do not do the intellectual spadework, who will? Support from DARPA or the National Institute of Standards and Technology (NIST) (see below) to develop a common architectural vision for future standards development may, at some point, be called for.

A healthy standards process helps foster good standards but, in the end, cannot guarantee them. Standards that prevail in the market necessarily reflect market forces: the desires of consumers; the strategies of players; and, yes, the role of public policy.

Chapter Six

THE PLACE OF STANDARDS

The administration's objectives for E-commerce include expanded markets, protected privacy, antitrust enforcement, fraud prevention, and the protection of intellectual property rights. Clinton and Gore (1997) cited standards as critical because

> they can allow products and services from different vendors to work together. They also encourage competition and reduce uncertainty in the global marketplace. Premature standardization, however, can "lock in" outdated technology. Standards also can be employed as de facto non-tariff trade barriers, to "lock out" non-indigenous businesses from a particular national market.

In theory, standards are a phenomenon of business that the government could easily stay well away from. They are very technical details of a technical enterprise. Thirty to forty years ago, when the information technology arena was smaller and less central to the overall economy, the federal role in and influence over the field was larger. Many federal users were leading-edge consumers. Then, an activist federal policy in developing standards made sense. Today, the government is but one user and not notably ahead of others.

Indeed, are standards a problem that demands government intervention? Arguing that a lack of standards is depressing *today's* growth curves is not easy. Business-to-consumer curves are growing nicely,[1] and market valuations of Internet companies are unprecedented. One might as well argue that "Internet time" is too slow.

[1]The growth curve of business-to-business E-commerce is less dramatic but still impressive. After all, such E-commerce has been under way for close to 20 years. There is thus a working model and a set of relationships that need only be electrified.

But cultivating a blissful ignorance of how standards are evolving and what such evolution may mean to the market is unwarranted. First, government policymaking will affect the standards process in any case. It helps to understand how standards work if deleterious effects of such policies are to be minimized. Second, standards have the potential to become a policy tool in their own right (and, if advocated as such by the United States in international forums, must be done in a well-informed manner). Third, there is an important role to be played by a neutral, third-party convener for standards.

THE PATENT TRAP

Generally speaking, the government has smiled upon standards groups, going so far as to devote a large share of NIST to fostering their success. Although antitrust objections might theoretically have been raised to thwart their work,[2] no such objections ensued in practice. The deliberations of the aforementioned focus group, however, indicated that no other obstacle so vexes the standards world as the growing specter of software patents.[3]

One focus group member argued that the

> patent situation is approaching the edge of insanity in terms of what can be patented and incompetence in terms of the lack of awareness about prior art and obviousness; claimants are coming out of the woodwork with patents perceived as trivial by many computer scientists . . . [they] are becoming a major threat to our ability to standardize, develop infrastructure for, or even advance information technology.

Another observed that the patent situation is intolerable and that standards groups are unable to do their work. He added that whatever difficulty being caused by actual patents has paled before the "fear, uncertainty, and doubt that has been interjected into the game," further noting that the prospects for litigation have created an incentive to write unclear patents that disguise the potential for

---

[2]Indeed, the Open Software Foundation, a UNIX standards group, was sued by Addamax on the theory that its deliberations reduced the market prospects for the plaintiff. But Addamax lost.

[3]Tim Berners-Lee, a focus group member, put his objections in print in Berners-Lee (1999), pp. 196–198.

violation until it is too late. Many of these patents were not novel but extensions from the physical world into the virtual. A third found that the patent situation has

> made it difficult to write standard disclosure and licensing agreements that will assure large companies that their ancillary patents will not be infringed upon [and] they are a [particular] barrier for small firms which lack both expertise and a large pool of patents of their own.

This concern was echoed by a fourth focus group member. Patents, in general, were perceived as an unnecessary burden on the process of standards formation—not so great as to stop something that everyone really needed (e.g., XML) but enough to halt progress on ancillary agreements (e.g., the Platform for Privacy Preferences Project [P3P]) that collectively fertilize the Web's rich ecology.

As one reaction, the IETF, which used to pull a proposed standard automatically if someone claimed patent rights on it, now lets working groups decide on their own. As another, the Federal Trade Commission enjoined Dell Computer from collecting license fees from users of a computer-bus standard that incorporated patentable technology, the existence of which was not revealed to standards writers until they finished. By contrast, the W3C lacks a policy on patenting technology developed jointly by its members. Some have thus used their patents to access other companies' technologies, as well as for license fees. For instance, without announcing its application, Microsoft received a patent for "cascading style sheets," which covers not only HTML but also XML. Microsoft has said that it will give away the patented technology for free, in exchange for access to patents by other companies—but has still upset its colleagues in the W3C, many of whose members have no declared intention of limiting the use of their patents only to ensure technology interchange. More notorious was a company that allegedly attended the deliberations of the P3P discussions while simultaneously preparing a patent (whose content was continually amended as discussions progressed) and springing it upon members as the P3P effort was concluding.

Patents may do more than interfere with standards development: They are two different ways of organizing markets for meeting new opportunities—and thus competition within such markets. How, for instance, should digital content, such as music, be protected from piracy? One proposal (the Secure Digital Music Initiative [SDMI]; see Appendix D) would encode music content in a standard form, around which music player software would be written. Another would use a proprietary formula (e.g., from InterTrust, an SDMI member) protected by patents—whose success or even presence may block or blunt the ability of a public standard to evolve toward greater functionality. How can reviews of products (e.g., books) be found, given that one document cannot easily point to another written later? One method may be a proprietary service using patented business process methods. Another may be through a standard way of tagging reviews to material so that a search engine can scoop up all the reviews. Take comparison shopping—such as rating colleges. Will it be a service hosted by a patent-rich Web site or, as Appendix E suggests, an application that any (shopping ro)bots can do if college characteristics are described in standard ways? True, standards and proprietary advantage are hardly antithetical. CommerceOne, a business-to-business E-commerce firm with a high market valuation (as of the end of 1999) was spun off from CommerceNet, a standards-promoting entity. Many standards (notably in telecommunications) have incorporated patents licensed on a "fair and reasonable" basis. But the Web's easy ubiquity may enable sites to exploit patents to obviate the need for standards (service is just a click away). Power to the patentee will give them a disproportionate top-down ability to plan the evolution of technology, even one built for an Internet that evolved from bottom-up experimentation. The result of this tension will tell how public the Web will be.

STANDARDS AS A POLICY TOOL

Standards may be able to take over much of the work otherwise required from politically controversial regulation—especially in the digital age. The evaluation of digital material can become easier and more automatic than the evaluation of written material (e.g., how many people read labels on canned goods?). Standards can be a mechanism that shifts oversight authority from government (if national governments are even the right level for a *World* Wide Web), not to overworked consumers, but into software whose parameters permit the careful consideration of categories rather than the rushed judgment over each case.

The role of voluntary standards as a substitute for regulation is epitomized by the voluntary rating system adopted in the 1960s (succeeding the Hollywood Code). It gave adults some hint about a movie's content and helped regulate what children saw. Although imperfect,[4] it seems to work. Parallel efforts for television combine content ratings with electronic locks (V-chips) that read the rating signal—hence the need for standards—and can screen out certain shows. With the Communications Decency Act declared unconstitutional, parental regulation of Internet content relies on purchasing nanny software, which ought to grow more effective if and when rating systems for Internet content mature. It remains to be seen whether potential ratings systems for *privacy* will put the issue to bed and avoid the need for a European-like solution.

Nevertheless, if software is to evaluate content, there must be standards by which content can be automatically recognized and its veracity vouched for (a point understood by TRUSTe, as noted in Appendix D; it not only writes standards of behavior for Web sites but monitors compliance against them). Perhaps third-party reviews (e.g., how one knows that a product is "safe and effective") can become an important part of the voluntary regulatory process. For this method to succeed, it has to be easy to find such reviews—a problem when links from review to product are straightforward, but links from product back to review are not. A standard set of back links (implemented in, e.g., XML) may permit search engines to list reviews of any well-labeled object without the usual clutter of extraneous material.

In enhancing security, notably infrastructure security, the federal government is uncomfortably seated between controversial regulation and ineffective hectoring. Standards by which institutions may have their security policies and practices rated by third parties might provide a lever for improvement. As ISO 9000 seems to have done for quality control and ISO 14000 for environmental management, the steady pressure of outside review, both good and bad, has a way of pushing people to adopt good practices, if only defensively.

---

[4]Witness the hybrid PG-13 and NC-17 ratings, the complaints that violence is accorded softer treatment than sex, and the irony that Hollywood feels that G-ratings dampen ticket sales.

If the government is to exploit standards as a substitute for policy, it may need to put some money into their development. A clear consensus of the focus group was that funding agencies (e.g., the National Science Foundation [NSF]) look kindly on researchers using part of their grant money to participate in the standards process. With open access to a well-structured universe of knowledge critical to its advancement, research and development on conceptual frameworks (e.g., ontologies) may also merit support.

Does standards advocacy need a standards advocate per se? The issues just noted—e.g., privacy, patents, and security—are debated at the cabinet and subcabinet levels. But the highest-ranking individual that deals with standards on a regular basis is the head of NIST's Information Technology Laboratory (ITL)—three levels below the Secretary of Commerce (via the Under Secretary for Technology Administration and the head of NIST). In national security affairs, elevating an issue entails using a chair in the National Security Council (NSC). There is no easy analog on the domestic side. Neither the National Economic Council nor the Office of Science and Technology Policy have the longevity of the NSC, and neither is involved in day-to-day management or closely linked to DoD (whose acquisition clout matters). Furthermore, raising an issue has a concreteness absent from raising a viewpoint that applies across issues, particularly one with such broad reach.

NIST'S EVOLVING ROLE

ITL has long been involved in the standards process, adding its technical expertise or good offices to promote standardization and representing the government's interest in this or that feature. To the extent that the government has sought an interoperability strategy (e.g., the DoD's Joint Technical Architecture), NIST has helped compile lists of relevant standards and profiles thereof.

After the 1994 elections, ITL expected criticism for spending tax dollars doing standards work that private firms could do and did.[5] So, it decided to focus on the development of public infrastructure: tests,

[5]References to ITL prior to 1995 are to its predecessor, the National Computer Systems Laboratory, renamed when it absorbed the Computing and Applied Mathematics Laboratory that year.

metrics, and corpora by which adherence to standards could be measured.[6] It was the general consensus of the focus group that even more metrology was needed. Indeed, it often happens that the technical quality of a standard cannot be truly known until efforts to write a test for it are under way.

A second area in which NIST may make a contribution is to develop (or at least evaluate) technology that may facilitate interoperability. Are there generalizable features of standards that make them easy to implement, flexible against unknown changes in requirements, and clear enough to minimize the amount of hand-massaging required to make two compliant applications interoperable? Are there ways of ensuring interoperability with lighter standards that do not have to specify as much, or, better yet, with translators and mediators that can dispense with many higher-level standards altogether? DARPA is funding some technology, and more sustained efforts may be merited. NIST itself can develop the parameters, corpora, tests, and testbeds that help measure the quality and fitness of ontologies and mediators.

A third area where NIST could play a role is in the development of a terrain map for E-commerce and knowledge-organization standards.[7] Battered by a blizzard of standards activities—especially in the realm of semantic standards and resulting tag sets—many such enterprises have little inkling of who is doing what to whom or, ultimately, where the real action is taking place. ITL could provide a neutral meeting ground for various efforts; it could also document the current status and plans of the various groups to look for possible gaps, overlap, and contradictions. Erecting a reference structure may help standards processes and reduce overall coordination costs by letting everyone know where they sit in the rapidly evolving universe.[8]

---

[6]A *corpora* is a sample (e.g., ten hours of telephone conversation) against which technologies (e.g., speech recognition) are tested. NIST's testing role involves testing for *per*formance as well as *con*formance.

[7]Much as NIST's Electrical and Electronics Laboratory helped Sematech.

[8]The line between a reference structure and a recommendation is thin but critical. In the mid-1990s, NIST created profiles of standards selected for having an SDO imprimatur. Nevertheless, NIST's corporate history—embodied in its defining controversy over battery-life extenders—has made it reluctant to evaluate specific products; it

FOR FURTHER RESEARCH

Both the patents issue and the uphill climb to semantic standards merit more thought.

Fundamentally, what *should* be patentable (e.g., software frameworks, business models)? How can greater awareness about prior art be introduced into the Patent and Trade Office? What standards of obviousness should be applied (e.g., are virtual metaphors for an existing physical construct per se patentable?)? Should the patent application and enforcement process be more transparent (i.e., to reduce "fear, uncertainty, and doubt")? How can incentives to write unclear patents be reduced? As for their effect on standards processes: Can better patent disclosure and licensing agreements be written? Can companies be asked to disclose patent applications as a precondition of joining standards efforts? Is there a way to get a quick read of how standards processes may be jeopardized by specific patent claims? How might strategies of patent makers and patent takers evolve within the standards arena?

The research on semantic standards presumes a public interest in timely standards that neither split the user community into disparate camps nor bias the marketplace in anyone's favor. How can progress toward such an end be measured? What are the precursors of success or failure? What indicates that forking is appropriate (e.g., different standards matched to different needs) or inappropriate (e.g., differences among standards that reduce interoperability more than they improve the fit)? What determines whether standards solve the right problems, and how can this be judged by creators, consolidators, and users of content? Finally, are there techniques by which the health of standards *processes* can be judged?

---

validates the capability of independent laboratories to do this. It is even more loathe to handicap winners in standards contests.

Chapter Seven

CONCLUSIONS

Will the digital economy be well-served by standards? So far, standards difficulties have proven no worse than a speed bump before technology's relentless march. And standards developers have showed considerable flexibility in finding forums in which to generate agreement. But two issues loom as potential roadblocks.

One is the deleterious influence of patents on the standards process—a problem that standard developers may find ways to mitigate for their own purposes but that, ultimately, has to be resolved outside the standards process.

The other is the challenge of semantic standards. Here, too, the root of the problem is no less than the long-standing difficulty of encapsulating the messy world of human affairs into a clean form suitable for computing machines (or, before 1950, for mathematics). Perhaps this problem will never be solved. Or, perhaps, men and women of good will can find their way to a standard set of good-enough resolutions. Or, just maybe, someone will develop an approach simple and satisfying enough to become a standard on its own—so much so that people looking back will wonder why the problem ever existed.

Appendix A

THE WEB AS WE KNOW IT

The history of the Web has been one in which the interests of commerce and those of knowledge have shared an uneasy relationship. Because the Internet, gloriously, can support a near-infinity of communities simultaneously, both worlds can coexist. But a similar profusion of information technology standards—each incompatible with the others—makes it hard for multiple communities to understand each other's content and thus interact (another great promise of the Internet). Ultimately, a single arena must exist in which to settle the various claims of alternative approaches to transmitting messages, structuring text, manipulating images, handling money, protecting privacy, and so on.

The rapid commercialization of the Web owes no small debt to the development and refinement of its key standards. Two of them, HTML and Java, are examined in this case study.[1] As a standard for content formatting, HTML has been a solid base from which to expand and augment with new technologies. It first gained momentum at the grassroots level; formal standardization happened only

[1]Two other standards, URL and HTTP, also played a key role in creating the Web. Both have evolved more slowly and with less controversy than HTML. The URL specification became a draft standard in December 1994 (RFC 1738) and, although revised in July 1997 (RFC 2368), remained short of a full standard as of mid-1999. HTTP (version 1.1) became a proposed standard in January 1997 and a full standard in June 1999. The major complaint about HTTP is that it was optimized to forward small chunks of text to the viewer and is relatively slow at shuttling the much larger (and more consistently formatted) files associated with multimedia. Streaming audio and video files, for instance, are more often handled through specialized off-browser players (e.g., RealAudio, a proprietary format).

later. Java's history, quite different, has been characterized from the outset by conflicts and, at best, spotty commercial success.

HTML: THE HYPERTEXT MARKUP LANGUAGE

HTML did not mark the invention of hypertext, a concept introduced in 1945 by Vannevar Bush and increasingly discussed by the academic community in the 1970s and 1980s. What HTML did was to wrench the focus of hypertext standardization from the problem of formatting documents of increasing complexity to that of supporting communities that could exchange straightforwardly structured documents among themselves. By piggybacking on Internet-based external addressing, HTML permitted authors to refer to and, more importantly, be referred to by the rest of the world.

HYPERTEXT BEFORE THE WEB

Text is a linear medium; hypertext is nonlinear: small fragments of text (and graphics) interconnected by machine-supported links. As such, the problem of representing it most closely fit the generic problem of representing compound documents: text with markup.

The reigning standard for representing markup is the Standard Generalized Markup Language (ISO 8879-1986), created by IBM's Charles Goldfarb. It enables an author to annotate a document in two ways. Material could be put between tags—<example> as this text is </example>—to denote a particular treatment of that text. Or material could be placed separate from the text, often a fragment of a database: e.g., <document author = "Vonnegut" category = "fiction">. Markup also provides de facto templates that ease parsing (and indexing) information for search and retrieval. Yet, markup, to make sense, must be accompanied by a document type definition (DTD), which specifies which tags are used, the structure, and the relationship to other tags (e.g., all chapters are composed of sections, which are themselves composed of paragraphs and pictures).

Charles Goldfarb also developed the circa-1990 primary standardization effort of hypertext, HyTime (ISO 10744-1992). HyTime reflected the problem of representing music as a set of synchronized events by supporting a lengthy vocabulary list of linkages and such.[2] For a while it looked as though the linkage features of HyTime could proliferate even if the timing and event-tracking technologies would require a difficult-to-implement second engine operating atop the SGML engine.[3]

Prior to the Web (and aside from Apple's Hypercard), simple hypertext was most often used in help files. The result is like an automobile manual: Users are prompted to choose among questions (e.g., does the engine have trouble starting?), each tagged to a different section of the manual, with yet more questions (e.g., what color are the engine deposits), and so on, culminating in suggestions for repair. Manuals are intended to be followed, not read. Used this way, hypertext had little need of standardization. Someone who needed hypertext material for, say, employee training would probably get the engine bundled with the hypertext and not look for more material except from the original vendor.

Hypertext, though, always had the potential of realizing another role, one envisioned by Vannevar Bush and Ted Nelson: linking text to *other* people's documents. Almost all text and imagery refer explicitly or implicitly to other text and imagery. The typical hard-copy article with footnotes and a bibliography requires the reader to find a way through a heterogeneous mass of material. Some authors assume imperiously that the reader should know the background. Successive newspaper articles on long-running stories repeat the background again and again. These techniques, while unsatisfactory, are accepted as limitations of physical media.

With digital media, accessible cross-references can directly refer to other text, pieces of text, or menus compiled by a third party. Once hypertext is used *externally*, however, it begs to have both its references and reference methods standardized so that they can be accessed even if held and managed by various parties. No one per-

---

[2]HyTime defines standard markup terms for the definition of hyperlinks (of which there are five: independent, property [simple element-attribute associations], contextual, aggregate-location, and span); the specification of coordinate addresses; activity-tracking indicators (e.g., when the section was modified); and references, both internal and external.

[3]Steve de Rose, founder of Electronic Book Technologies, interviewed on November 10, 1992.

son can index and cross-reference the annual output of 300,000 medical articles, yet experts do it using overlapping fields of expertise. It suffices that references be consistent and that the links be presented uniformly.

The World Wide Web Is Born

Circa 1990, the stage was set for combining the need for *external* hypertext with the Internet's growing tendency to connect everyone. That year, a small group of researchers led by Tim Berners-Lee at CERN (the European particle physics laboratory) developed the concept of the World Wide Web. The implementation of this concept involved several elements, including ways to address a computer's *files* on a network, the first version of the document structure standard known as HTML, and a "browser" program capable of displaying documents on the Web with a computer. HTML, a subset of SGML, provided a syntax for document structure independent of its presentation and formatting on any particular computer. Indeed, the first version of HTML had *no* provisions for formatting and presenting information. Leaving this task to the user was essential to solving the problem of portability between computers of different platforms attached to the same network. HTML provides a straightforward means of accessing content on other machines, by bootstrapping on the DNS—already established on the Internet for transferring bits—which references computers by using a hierarchy of physically connected networks. Individual HTML files resident on these computers are specified by their "local" filenames within the hierarchy of files and directories on that system. Addressing material by location, rather than by content, or by a persistent object identifier (see Appendix C) was an important architectural feature that permitted rapid usage but has since led to many structural problems (notably dead links—the dreaded "404" message).

From 1990 forward, HTML *was* the Web. Its simplicity enabled the conversion of large amounts of data to a Web-shareable format by all, backed, as they were, with easy-to-use tools. HTML, in combination with the Web, exploited the access to other users that networks provided, allowing the author flexibility in the extent and manner of reference to other knowledge. The fundamental usefulness of HTML plus HTTP was its ability to present a document at the same time it was being retrieved. Hitherto, file transfers (via FTP: file transfer protocol) required guessing at a document's contents by its title, downloading it, reading it with other software, and then returning to the file directory if the document was unsatisfactory.

HTML initially defined a small set of tags:

- **Anchors**: The <A> tag referred to another document on the Web by specifying a Universal Resource Locator (URL). It represents one of the most important innovations of HTML, one absent from SGML.
- **Titles**: <TITLE> and </TITLE> tags provided for this most basic function, also creating a clearly identifiable point for summarizing the content of a document.
- **Lists**: Ordered lists (using the tag <OL>) and unordered lists (using <UL>) permitted the delineation of individual items. Line indentation (<LI>) let lists stand out as such.
- **Headings**: Six levels of header tags (<H1> through <H6>) could be used to mark chapters, sections, or other structural cues in a document. Taken together, headers could be separated from the rest of the text in the document to understand structure.
- **Paragraphs**: The <P> tag delimited paragraphs.
- **Fonts**: Support was provided for underlining (e.g., <u>text</u>), bolding, italicizing, and blinking text. A monotype font permitted the orderly arrangement of tables and ASCII-level illustrations.

And that was it—enough to format text and link into related documents. The output was not very fancy at the time, and the level of representation was a few years behind the state of the art in word processors (e.g., Word Perfect 4.2). But anyone could master the rules and start throwing material onto the World Wide Web, literally, within hours. From there documents could go anywhere.

In 1991, the CERN researchers started to discuss the concept and features of HTML via E-mail discussion lists. As the user population of technologists grew, so did the debate about the next set of HTML features. With so many diverse applications envisioned for the Web, it was not surprising that different groups formed different, and only modestly compatible, versions of HTML for different functions. To view HTML-tagged content correctly, people had to use browsers that recognized these tags. Web users—still few—showed their support for a particular HTML style by playing with various browsers displaying various types of content and responding directly to their creators. The authors of browsers and content knew exactly how good they were by seeing who read what they had to say, who used which browser, and what improvements they suggested. Chaotic interactions nevertheless crystallized a consensus not only about the form of HTML but the potential of the Web in general.

By April 1993, two notable browsers existed: a text-based browser called Lynx (written by Lou Montulli) and a graphical browser called Mosaic (developed by Marc Andreessen, et al., at the National Center for Supercomputing Applications [NCSA] at the University of Illinois). A de facto HTML standard could still be defined as the features supported by both browsers and in general use among Web page writers. But the graphics features of Mosaic—the pretty pictures—were what spurred the Web's proliferation (prior to which it was considered a library research tool[4]). Among the multiple standards for representing images at various levels of compression, Mosaic chose Compuserve GIF, standardized in 1987 and 1989. It was easy to decompress, fast to display, and offered a modest level of compression. Users could pull some very striking images from cyberspace. The National Aeronautics and Space Administration (NASA), in particular, had a very large stock it was willing to share with an eager public.

In May 1994, at the first World Wide Web conference in Geneva, participants presented a variety of ideas for a standardized version of HTML, known as HTML 2.0. The IETF simultaneously sought to standardize HTML by establishing a working group open to anyone who understood the issues. In July 1994, the group issued a specification for HTML 2.0 (Request for Comment [RFC] 1866), which collected the most commonly used tags.

A few months later, as the commercial potential of the Internet became more apparent, the W3C was formed to develop Web stan-

[4]Ed Krol's *The Whole Internet* (1992) classified the nascent World Wide Web as a "research librarian" tool together with Gopher and WAIS (which is discussed in Appendix C).

dards. The W3C differed significantly in makeup from the IETF by including representatives from the major corporations that would develop Internet products (e.g., Digital Equipment, Hewlett-Packard, IBM, Microsoft, Netscape, and Sun).

In March 1995, a draft of the HTML 3.0 specification reached the IETF. (See Galbraith and Galbraith, 1998.) It included many controversial extensions, such as ways to wrap text around images. The most controversial dealt with tables. Several representatives from government and industry raised concerns about how tables should be implemented; all wanted to support their own applications and minimize their own production costs.

As commercial interest in providing products and content for the Internet grew, so did the volume of debate over HTML's features. The more seriously companies and individuals thought about how to produce and convey Web content, the more HTML standards debates focused on formatting and presentation issues and the less about the structural aspects of the standard. As with other media, commerce had begun to edge knowledge from the standards arena as the virtues of community consensus yielded to the dictates of de facto standardization through commercial products.

And Then the Browser Wars Began

In late 1994, Marc Andreessen left the NCSA to cofound Netscape Communications with Jim Clark. (See Netscape, 1999.) Through 1995, Netscape released beta versions of its new browser as a way of building a mass market. It also introduced some new HTML tags in its browser and began discussing refinements to its browser (and HTML) widely in the user community. The addition of some formatting and presentation tags to browsers during this period allowed HTML content authors to specify colors, font sizes, and other aspects of the appearance of their documents. Such extensions to the HTML 2.0 tag set exceeded the standards but provided innovative ideas that served as trial balloons for user feedback.

In November 1995, Microsoft introduced the Internet Explorer browser. In response, Netscape revised its own browser to include support for Sun Microsystems' Java (see below), as well as frames, a new HTML construct. (See Netscape, 1999.) To help organize complicated documents, frames split the browser's viewing window into sections: For example, a book's table of contents could be placed in one frame and its chapter texts in another, so that scrolling the content would not take the table off the screen.

Having two major browsers, each with incompatible extensions to the HTML standard, dislocated the standards process well beyond the scope of previous debates. Tables and frames that looked correct when formulated in one browser worked poorly when read in another. With the de facto standard rapidly falling under the domain of the commercial browser makers, the IETF's HTML working group found itself overwhelmed with suggestions and complaints regarding its standard. So, it disestablished itself. The W3C, for its part, formed an HTML Editorial Review Board, which represented a small number of companies (e.g., IBM, Microsoft, Netscape, Novell, and Softquad). By including only a few large commercial stakeholders and a few technical experts in the development of the standard, the board was able to strike a number of agreements on HTML. They added neutrally implemented tag formats that both browsers already supported and deleted tags, such as <BLINK> or <MARQUEE>, that were perceived as superfluous or offensive. Agreement was reached in January 1997 on what became known as HTML 3.2. The new standard provided support for tables, applets, and text wrapping around images.

HTML 4.0 (December 1997) extended HTML 3.2 by supporting enhanced tables, complex documents, in-line multimedia content, formatting control in style sheets, printers, international languages, and methods by which the disabled could access documents. (See W3C, 1999.) As with version 3.2, HTML 4.0 essentially just codified the most popular features supported by both browsers. In May 1998, the W3C stated that HTML had reached a degree of maturity, and HTML work was superseded by work on XML, discussed in Appendix B.

In the end, the standardization efforts over HTML, brief as they were, had an ironic effect on the software industry. By 1995, Microsoft had already become its top gun, with both the lion's share of the operating system market and the primary office application market. Those tempted to use competing (and often cheaper) applications were daunted by the fact that documents formatted by one suite (be they text, graphics, or spreadsheets) translated unreliably to another. Dominance was extended by technologies (e.g., object linking and embedding [OLE]) that permitted applications in one area (e.g., Word) to exchange data with applications in another (e.g., PowerPoint, Excel)—making translation even further fraught with peril. In 1995, Microsoft concluded that the browser was not just another product (and one often given away, at that). Rivals had boasted that the operating system would pale into insignificance beside the browser as the entrée into the world of documents (a claim that continues to this day). Initially, as noted, the two browser producers competed by encouraging the authors to use features that read well in one browser but not another. Had this continued, the world of Web pages might have been split, with the probable result that Microsoft's would create a bandwagon that would consign Netscape to a niche. But with HTML standardization so well established, such a strategy had scant hope of succeeding; so Microsoft tried more muscular tactics. These tactics, in turn, attracted the attention of the U.S. Justice Department.

JAVA

Java (né Oak) was originally developed in 1991 as a language to manage the "electronic home"—networks of household appliances, different enough to require a controller of unique size and complexity to run it. Java's design, at that stage, was constantly being redirected to respond to fleeting commercialization opportunities for this or that appliance.

Once the Web's potential became apparent, Sun repositioned Java for that market, jump-starting its bandwagon. With incorporation in the Navigator 2.0 Web browser, Java received major play and a nearly instantaneous user base on the Web.

Java technology has three components. (See Kramer, 1996.) One is the Java Runtime Environment, which layers a Java Virtual Machine (JVM) between the operating system and the applets written in the Java language. Two is the programming language in which applets are written. Rather than being compiled as a stand-alone executable file (e.g., ".exe") for a specific platform, an applet is interpreted by the JVM. Code written in Java could be downloaded by users running any computer, regardless of make or operating system, as long as it ran a JVM-enabled browser. Three is the predefined functional class libraries (referenced through the Java Native Interface, an application portability interface [API]) which are constructed in the course of running the applet. Combined, they permit the creation of applets that can be run by the operating system, with some safety: Java's security policy makes it hard for applets sent over the Internet to access vital low-level system resources, such as a computer's hard drive.

Applets can be integrated into Web page content as HTML extensions. Web designers could provide extra functionality at a rudimentary level (calculators, data entry tools, etc.) to enhance Web pages. Applets run slowly because they must be repeatedly interpreted within a browser on top of an operating system, all of which adds overhead. But the de facto standardization of Java allowed developers to get up to speed quickly at writing Java code and thus access the majority of users with Web browsers using Netscape Navigator at the time.

Java allowed Web site designers to put real functionality in Web pages. By embedding an applet in a Web page of, say, a map, a user could zoom to various levels, and turn map features on and off without having to leave the Web page to use a spreadsheet or other application (retaining greater control over the user's visual environment mattered to site producers). The map could be displayed side by side with other information on the Web page written in HTML, such as the traveler's itinerary. When introduced in 1995, Java permitted far more control over graphical appearance and dynamic data than did HTML 2.0. Because all browsers ran Java, Sun won over developers otherwise wary of devoting their efforts to learning yet another computer language. The limited functionality of applets might have been inadequate for some applications, but it at least got programmers interested.

The Java Strategy

By combining browser technology (which permitted files to be accessed from anywhere) and software to run Java (whose applets could be run anywhere), Sun and Netscape hoped to diminish the effective utility of any specific hardware-and-operating system architecture, thereby freeing programmers from having to concentrate on one at the expense of others. Java was part of the strategy to reduce the role of the Windows operating system, relegating it to a commodity product and, as such, removed from its pedestal. Then Sun and Netscape could insert themselves as the primary value-added component in the computer system.

Sun realized more success in having Java accepted as a multipurpose programming language than as an applet language. As an object-oriented programming language, Java is more constrained than C++ (it imposes resource management constraints, such as mandatory garbage collection and other memory management restrictions) but is less structured than Ada or Pascal. As all object-oriented languages do, Java encourages programmers to construct class libraries for commonly used functions. As a structured programming language with a unique technology, Java was a popular course subject at many universities. This created a base of young programmers to fill the perceived market for Java applications. By mid-1999, there were over 700 books in print on Java.

To programmers, Sun emphasized Java's technical merits: portability, security, technical novelties, and close ties to the Web. Portability was the strong suit. Until one operating system dominates all computers from palmtops to "big iron," programmers would have to cut into precious creative time to adapt their output to the unique architecture features of rival systems. Sun persuaded developers, contractors, and university professors to invest up front, thereby circulating Java literature and developer kits within the software community.

To development managers of large projects, Sun emphasized Java's rigorously structured syntax and code library management strengths, highlighting the economics of developing code only once for use on multiple platforms. Corel, which owned Word Perfect, even rewrote applications in Java (but they never reached market). Perhaps more importantly, Sun and venture capitalists funded start-ups dedicated to developing various enterprise applications for Java to take advantage of the evolving networked corporate environment. Many such products, notably those that aided server operations, became commercialized.

But Java, despite its apparent ubiquity on the Web, fell shy of its expected trajectory. Improvements in browsers obviated many of the advantages of Java. Developers soon realized that the limitations of HTML could be overcome if there were support for several functions in later browsers. Navigator 2.0, for instance, built animated

GIF images—simple multiframe pseudoanimated content—into HTML. This was a boon for advertisers, depending, as they did, on appealing to user eyeballs. With advertising being the major revenue model at the time, this was a big deal. Animated GIFs were easy to develop, and they ran with far fewer resources than applets required.

Netscape's Navigator team also developed a scripting technology for Web pages named Javascript (né Livescript until Netscape allied with Sun to distribute Java with its browser). Its purpose was to augment HTML's structural orientation with presentation features, such as background colors and graphics and the ability to create simple graphical interfaces and related utilities. Javascript shares its syntax with Java, but little else—notably technology. Javascript's rapid uptake was due to its easy implementation. Microsoft, in response, released its own scripting technology, ActiveX, which, although not without users, was never as popular as Javascript. By 1997, Microsoft acceded to the market and included Javascript in its Internet Explorer; ActiveX was repositioned to support its Common Object Model (COM). Netscape eventually petitioned ECMA to accept JavaScript as a formal standard. The result, ECMAscript (ECMA-262), is almost identical at the functional level to Javascript. (See ECMA, 1999.) By contrast, Java's complex design was not altogether justified by performance benefits; in some sense, it was a solution in search of a problem. This inhibited its use on the more popular commercial Web sites.

Having been tried out with less than total success on the Web, Java returned to its roots. Sun tried to reposition Java as an operating system for lightweight platforms ranging from computers lacking much memory or storage to appliances. Several "thin client" products based on Java were tested, but they were technically limited. The plunge of personal computer (PC) prices below $1,000 in late 1997 cut the cost differential between them and "thin client" machines. High switching costs and the absence of a compelling reason to move beyond the personal computer put paid to the Java platform. It would take the long-predicted move toward pervasive, networked appliances before Java could be reignited as an operating system. Sun's subsequent Jini initiative now seeks to augment the core Java language with protocols for networked appliances. With Jini, Sun hopes to attract developers and vendors to the broader Java platform. The forthcoming market battle for the soul of the networked appliance may come to pit Sun against Microsoft, 3Com, and others.

Java's Formal Standardization

Popularity results in de facto standardization, and Sun's ability to standardize Java met various levels of success in its three markets: as an HTML extension, as an operating system, and as a programming language. But Sun also pursued de jure standardization.

In 1997, with Java established among programmers and still a buzz in the market, Sun sought ISO imprimatur for it. A broad standards organization, ISO had global and cross-industry recognition for both general standards (e.g., ISO 9000 for auditing quality control processes) and computer language standards (e.g., FORTRAN, Ada, and C++). But, typically, ISO standardization was a slow process. It took C++ over a dozen years to travel from first release in 1985 to standardization.

So, Sun tried a different tack. Several years ago, ISO developed the Publicly Available Specification (PAS) process, which differs from its usual Joint Technical Committee (JTC) process by allowing other entities, including for-profit corporations, to submit specifications for straight ratification votes. The PAS process also permits the submitter to make substantive changes in the standard without undergoing the formal JTC process.

Sun applied for approval as the sole PAS submitter for Java in March 1997. With 8 ayes, 15 nays, and one abstention, Sun was rebuffed and told to address three points.[5] First, it needed to define what Java included and what technology products would be submitted with the PAS application. The U.S. delegation, for instance, argued that

> In order to properly assess the qualifications of a candidate for Submitter of Publicly Available Specifications, it is necessary to be able to assess those qualifications in a specific context. In the case of a single for-profit company, the scope of the company's business

[5]See ISO/IEC JTC1 N4811 and N4833 (Summary of Voting on Document JTC1 N4615, Application from Sun Microsystems, Inc. for Recognition as a Submitter of Publicly Available Specifications for Sun's JavaTM Technologies).

> may be very broad or unclear. Where the scope of proposed submissions is not obvious by the nature of the company, the scope of potential submissions should be explicitly identified. SMI has included a scope statement in an amendment to its application (ISO/IEC JTC 1 N4669), indicating its willingness to identify specific scope. The U.S. seeks confirmation that only those technologies listed within the scope statement will be considered for PAS submission under this PAS application. To enable appropriate evaluation, the scope statement should identify the specifications proposed for submission by their current document references.

Second, most members cited trademark and patent claims associated with the use and licensing of the Java platform as a barrier. Because Sun makes money by licensing JVMs and the right to the Java trademark, most members raised concerns about Sun's ability to lock in market share through licensing. Many wanted Sun's reassurance that it would comply with standing ISO patent policy with its nondiscriminatory licensing arrangements. Sun's position on arranging trademark licensing and use also needed clarification.

Third, there were concerns over how the standard was to be controlled and maintained. Few were satisfied by Sun's assertion that changes and additions to the standard would be driven by its own "reasonable processes for broad consensus." JTC members felt that ISO should have that responsibility. Such comments reflected the JTC's general reluctance to approve a private company as a PAS submitter.

Sun responded with clarifications. In scoping the Java PAS, Sun asserted that the technology covered would include, "the JVM, the Java language, and the Java class libraries [APIs]." Sun further agreed to abide by the ISO's nondiscriminatory patent licensing policy and made it clear that it granted ISO fair use of the Java name for the general properties of the technology and the PAS specification. But Sun strictly limited the use of the trademark name for tag lines and specific products that used Java technology. Sun also wanted to maintain exclusive control of proposed changes and additions, relegating ISO to hosting an up-or-down vote. Finally, Sun offered to fund ISO's maintenance activities for Java.

That November, Sun was approved as a PAS submitter, with 20 ayes, 2 nays (including the U.S. delegation), and 2 abstentions. Scoping and IPR issues were resolved, but it was left to subsequent negotiations to determine what process would govern changes and additions—issues that mattered to the business strategies of those who opposed Sun's controlling Java's potential extensions. Such firms as Microsoft, Hewlett-Packard, and IBM wanted to compete with Sun by selling JVMs with potential platform-specific extensions designed for speed and perhaps at lower prices. They wanted the ability both to use Sun's trademarked Java platform and still to extend and modify the standard.

Success at ISO was dimmed by later difficulties in working out details of the process for changing and maintaining the standard. Sun was accused of wanting to retain unilateral control over the content of the standard and the selection of JTC members. Sun countered that competitors were blocking technology development. One weakness of the PAS process is that it deferred these issues beyond the original approval-and-disapproval stage. So, in early 1999, Sun declared it would submit the Java language, Virtual Machine, and class libraries to ECMA for approval, hoping to leverage the fact that ISO authorized ECMA to put standards on a fast track for ISO approval.

Sun has also defended the Java standard in the courts, suing Microsoft in 1997 for violating its Java technology licensing agreement. Microsoft's extensions to Java's class libraries and JVM, it argued, violated its agreement with Sun. Microsoft was accused of trying to leverage its near-monopoly of PC operating systems to extend Java and do so in ways that enlarged Microsoft's influence in markets where the demand for Java technologies might develop. Microsoft, in response, held that its extensions were meant to enhance functionality and improve performance and thus did not violate its license. In May 1999, a U.S. federal court issued a preliminary ruling in favor of Sun, forcing Microsoft to include Sun's Java API with its products. Microsoft announced it would appeal. Although the ruling would seem to keep Microsoft from including proprietary extensions to the Java API, the significance of this result remains uncertain.

CONCLUSIONS

The standardization of the Web is yet one more illustration of the power of strong and simple ideas to spread through their appeal and not through market or regulatory power. HTML and HTTP piggybacked upon established norms (e.g., markup languages) and institutions (e.g., the Internet). As both became accepted, they grew in size and complexity. Java was developed for lightweight devices, a market that is still developing. So Java was adapted to the Web, which, ironically found it somewhat heavy on top of the existing overhead of operating systems and Web browsers. Yet the hype attached to its introduction permitted a largely unrelated technology, Javascript, to launch from under its wings to reach great height. Java itself is returning to the world of the appliance rather than the full-fledged personal computer.

The Web has been fortunate in that both HTML/HTTP and Java were born as unique (and uniquely simple) ideas rather than as competing ways of implementing old ideas. The early rise of both permitted de facto standardization and signaled risk-averse consumers that the technology has matured (thus boosting sales). This hastens market consolidation down to a few vendors. Thereafter, whatever conditions encourage further standards-setting do so incrementally and slowly. As markets mature, so do standards. In the case of HTML and Java, the market has made the major decisions on adoption, leaving it to standards developers and corporate interests to jostle over the minor details of implementation.

Appendix B

THE EXTENSIBLE MARKUP LANGUAGE

Commerce is buying and selling. E-commerce is commerce with human interaction replaced by digital information insofar as practical. To work, it needs a commonly understood modality of exchange; a reliable description of the product to be purchased; and, in some cases (see Appendix E), a succinct statement of the buyer's expectations. People do all this by talking to each other, using brainpower, social cues, and a shared cultural context to figure out what each is trying to say. Machines, to repeat a familiar refrain, have only symbolic notation to go by. Thus, if E-commerce is to get past its current incarnation as mail-order but with keyboards, rather than phones, such notations must be explicit, mutually understood, and well-formatted—hence, standardized.

In a field that insists on shedding its skin as often as the Web does, it may be premature to say that the search for such a standard has ended. But today's bettors seem increasingly inclined to place their money on a metalanguage, XML. The use of *metalanguage* is deliberate. It has been remarked that XML solves how we are going to talk to each other, but we still need to agree on what we are going to talk about. XML is the grammar, not the words—necessary, but by no means sufficient. And therein lies both the hope and the hype of what may be the keystone of tomorrow's E-commerce.

In analyzing XML, this case study attempts to do several things: explain the broader advantages of markup, trace the history of XML through its origins in earlier standards, limn its current status, and portray the hurdles it must overcome to fulfill its promise.

XML AS A MARKUP LANGUAGE

The many limitations of HTML have prompted the industry to conclude that it is time to move beyond it. While HTML is a useful way to present information, it does little to organize it.[1] As one result, every time a document format is changed, the markup has to be redone, and a developer cannot alter a document's presentation without creating a different version of the document with new markup. XML tags, such as <price>5.95</price>, by contrast, indicate variables and their values when attached to any data element in a document. This permits XML documents to be processed by computers as well as read by humans. Tags can be used to drive searches and comparisons of data elements within a company's site, or across data sets provided by many companies. They also allow data to be arranged in specific ways for specific users. It provides users only the data elements that are of interest to them. In contrast, HTML tags contain only format information, forcing search engines to do textual analysis of Web pages and leading to many useless "hits" that do not fit the context of the search.

The strength of XML is that the standard opens itself up to an infinity of tags representing the infinity of objects and qualities one might want to keep tabs on. Unlike HTML, which has a single standard set of tags, however, XML tags can be defined document by document, application by application, industry by industry, or globally. (Data Interchange Standards Association, 1998.) This extensibility is one of the most attractive features to many users.

But it is also a potential weakness. For tags to permit cross-company or cross-industry comparisons, they must represent common concepts in commonly denoted ways. When people talk about the vast new global E-commerce markets facilitated by XML, they are

---

[1]HTML has been able to accommodate random parenthetical material and presentation hints since version 2.0. The META construct permits variables and values to be inserted as markups in documents. Some META constructs send information about an HTTP header field to an HTTP server; e.g., <META HTTP-EQUIV="Expires" CONTENT="Tue, 04 Dec 1993 21:29:02 GMT">. Other constructs are user-supplied; e.g., <META NAME = "television character" VALUE = "Barney">. In the META tag one can glimpse a early version of open markup but one that did not become the basis of a stronger descriptive vocabulary. The META construct lacked a way of defining a tag in a document (either directly or by reference), any structure, or any way to mark up text using such tags.

implicitly assuming that companies will create their catalogs and other information using common tags. Ideally, this would happen, and companies would compete on the qualities of their product and service offerings. However, the world is far from ideal. "Browser wars" between Netscape and Microsoft were fought in part through the use of proprietary HTML tags, as the two companies tried to expand the capabilities of their products to create better-looking documents and attract developers and users. Still, despite the possibility that some companies will not use standard tags, there are many efforts under way to agree on common tags within and across domains.

XML documents specify their tags and the relationship among them by leading off with or at least referring to a DTD,[2] which specifies what elements may exist where, what attributes elements may have, what elements must be found inside other elements, how elements may combine, and in what order. DTDs allow a validating XML parser (i.e., a computer program that reads XML), to determine whether the document's tags are "legal" and properly arranged for a given type of document. Those that fail generate error messages. But with every new DTD, a new set of tags becomes possible—hence the "extensible" in XML, a capability that HTML lacks; if a tag is not in the HTML standard, an author cannot define it into existence.

Forcing a tag to be a standard had its uses in HTML. If the browser recognizes the tag, it knows how to present the tagged information (e.g., whether to highlight, italicize, or offset the text). But Web designers found that they needed to use tricks to overcome the limitations imposed by the limited number of HTML tags. Some proprietary tags have been invented, reducing cross-platform useability (e.g., "This document best viewed with Netscape Navigator."). HTML has been used in ways that were never intended: single-point GIFs and too many tables. After a while, documents become hard to manage.

---

[2]Although most XML documents published contain DTDs, documents without DTDs may still be valid (by contrast, SGML requires a DTD for every document). An author can, in effect, create a DTD by implication—arranging tags in a way that an XML parser finds acceptable. Without a DTD, though, there is no automatic way to check whether all the tags that should be present are present and tags that should be absent are absent.

XML tags for their part have few or no inherent clues for presentation. This forces the use of style sheets to determine how a document would be presented. The separation between content markup and style definition allows the same document to be processed or published without additional work. The ability to attach many style sheets to the same XML document allows much finer control of the way a document looks in various presentations without affecting content in any way. A designer may, for instance, use one style sheet for display on regular computer screens; a different one for small screens, such as Palm Pilots; a third one for display on browsers with graphics turned off; a fourth one for printing; etc. Each style can be defined to the satisfaction of the designer without requiring that a document's markup be redone. By contrast, such presentation control can only be achieved with HTML by creating the same data with different markups for each presentation, storing all these documents in a database, querying the requesting device as to its type, and returning from the database the version that matches the specific type of browser. Dynamic HTML uses scripting to display and redisplay pages, based on user actions. XML's ability to let designers use standard style sheets or to create new style sheets from standard components offers a tremendous economy of effort in separating content from presentation. In addition to economizing effort, style sheets are more varied and much more flexible than HTML tags.

Style sheet standardization is dealt with via the Extensible Stylesheet Language (XSL), a descendant of the Document Style Semantics and Specification Language (DSSSL—ISO 10199) with roughly the same relationship as that of XML to SGML.[3] The W3C published the first draft of the XSL specification on August 18, 1998. (W3C, 1998b.) Later versions have already been published, with the latest version published in conjunction with other specifications touching on XSL and XML. (W3C, 1998c.)

In addition to tag structures, XML also provides facilities for link structures (Cover, 2000) via the XML Linking Language (XLL), with its two major components: XLink and XPointer. XLink

---

[3]The W3C's proposed recommendation for "Associating Stylesheets with XML Documents" was released in late April 1999. To kick-start the creation of XSL style sheets, Sun Microsystems and Adobe sponsored a contest with prizes valued at $90,000 for those who could develop layout engines for Mozilla, Netscape's open-source browser software. See also Johnson (1999).

> specifies constructs that may be inserted into XML resources to describe links between objects. A link, as the term is used here, is an explicit relationship between two or more data objects or portions of data objects. XLink uses XML syntax to create structures that go beyond the simple unidirectional hardwired hyperlinks of today's HTML to include sophisticated multi-ended and typed links[4]

They include:

- multidirectional links (so that users can return to the original location via a corresponding link at the first link's destination)
- multiple-destination links (giving users a choice)
- links to fragments[5]
- link databases to store links (thereby making it easier to adjust to changing link addresses).

The XPointer language defines "constructs that support addressing into the internal structures of XML documents. In particular, it provides for specific reference to elements, character strings, and other parts of XML documents."[6]

GETTING TO XML

What was the point of trimming SGML to get to XML? SGML is a heavyweight language meant to tackle "large, long-term document publishing" (see Jellifee, 1998), such as DoD's entire corpus of technical documents. Yet, its very size and complexity made it "just too hairy for real people to get into; you could crack great big problems, but sometimes not do the simple things simply. Then the Web came

---

[4]See the W3C working draft on the XML Linking Language (XLink)(W3C, 1998b). Note also the comparison with the complex links of HyTime.

[5]With XLL designers can (1) place content directly into the document being viewed without user intervention (so that a document on, for instance, chemical compounds could be viewed and a section on fructose automatically inserted from an entirely different Web site), and (2) replace content in line with updated content from another document. Yet, if the original text has original markup that conflicts with the markup of the document being viewed, strange-looking documents may result. Also, direct insertion prevents the quoting author from adding his own markup for emphasis.

[6]See the W3C working draft on the XML Pointer Language (XPointer)(W3C, 1998a).

along and showed the power of doing simple things simply."[7] XML is designed for doing "efficient, small, short-term documents." (See Graphic Communications Association, 1999.)

Die-hard SGML advocates could have justifiably argued that the SGML's bells and whistles were no real barrier to designers—who could have simply ignored those features they did not feel like exploiting. But those who wrote programs (such as browsers) to read marked up text would have had to accommodate any feature that the text's authors felt like putting in. Within a delimited universe (e.g., defense contractors, automakers) this problem could be avoided by developing master DTDs that avoided the more obscure features of SGML. But once the challenge became interpreting random text produced by someone outside the institutional aegis, ignoring obscure features could easily have led to disappointment or disaster were such features to be used. This is an example of how the role and thus the content of standards designed to unify a heterogeneous corporate infrastructure under a single authority (e.g., CORBA) failed to fit the model of a Web that may encompass literally anyone.

Although the itch to lighten SGML was long-standing, only in mid-1996 did Jon Bosak of Sun Microsystems convince the W3C to create a working group for SGML on the Web. The SGML Editorial Review Board included chief information officers, Internet IPO architects, and standards editors. The original idea was to "put in everything that's proven to work . . . and throw the rest out." Within a year it had become the XML Working Group. Although the SGML community leapt on board instantly, the "Webheads" held off.[8] As Jean Paoli of Microsoft observed, HTML was a more-or-less standard, widely used tool that worked. By contrast, XML's early fans were those least happy with HTML's limited power. (Seybold Publications and O'Reilly Associates, 1997.) Once Microsoft decided to use XML in its Channel Definition Format (its "push" technology) and announced the decision in March 1997, XML began to generate significant interest among programmers and Internet professionals. (Seybold Publications and O'Reilly Associates, 1997.) XML has been in ascendance ever since.

---

[7]Tim Bray, interviewed in Veen (1997), p. 1.

[8]Tim Bray, interviewed in Veen (1997), p. 2..

XML, by restricting choices present in SGML, grew simpler (see Johnson, 1999):

- A specific choice of syntax characters was made so that everyone using XML will use the same concrete syntax. For example all tags must begin with "<" and end with ">". Attribute values must be enclosed in quotes.
- A new empty-element tag was invented to indicate that an end tag is not expected. It looks like this: <some text/>.
- Tag omission is forbidden; each nonempty element must have both a start tag and an end tag. All tags must be properly nested (e.g., this is <b><i>wrong</b></i>).
- DTDs may be omitted.

Nevertheless, XML is not *that* much lighter. After all, every legal XML document is also, by definition, a legal SGML document. SGML has had a hard slog in the marketplace, accepted only in some communities. So why the optimism for XML? HTML helped; it taught both professionals and amateurs the value of working with markup. With HTML accepted, XML is seen as a way to overcome the limitations facing HTML. Users have moved past presenting pages and are looking for capabilities to search, collate, and move information and to allow computer systems to communicate without human intervention. Proclamations and product announcements by mainstream Web firms, such as Sun, IBM, Lotus, Oracle, Adobe, and Microsoft, have raised the odds that XML could become central to the Web's future. (Alshuler, 1999.)

If the purpose of XML was "to enable generic SGML to be served, received, and processed on the Web in the way that is now possible with HTML," the recasting of HTML into XML format has to be key. On May 5, 1999, the W3C HTML Working Group released a revised version of "XHTML 1.0: The Extensible HyperText Markup Language. A Reformulation of HTML 4.0 in XML 1.0" (W3C, 1999a) which provided a new set of modularized XML DTDs for HTML. By breaking up XHTML into a series of smaller element sets, it permitted the combining of elements to suit the needs of different communities. How easy or smooth will the transition from HTML to XHTML be? XML has much stricter rules than HTML, and XHTML is expected to comply with the rules of the XML specification. The key is whether programmers who are used to playing fast and loose with current HTML are willing to trade that for the greater expressive power of XML.

In general, SGML's advocates have been helpful to XML. Until XML came along, the largest single source of support for SGML was DoD's logistics community, whose CALS (né Computer-Assisted Logistics Support) program imposed requirements on defense contractors to document their technical support material in a standard way. Text was to be rendered in SGML, images in a series of ever-more sophisticated standards culminating in STEP. Groups developing STEP (largely in the aerospace and automotive sectors) realized that they can use SGML, and now XML, to integrate product documentation fully into product data management, to view structured information repositories of complex documentation and legacy data warehouses via Web browsers, and to manage technical and administrative flows of information within supply chains and consortia. (Wrightson, 1999.) As a result, efforts for full harmonization between STEP and SGML/XML are under way. The U.S. government's CALS standard has officially shifted SGML to XML. Meanwhile, NIST is transferring resources from three-dimensional representation (Virtual Reality Modeling Language [VRML]) into XML. The Text Encoding Initiative, a project funded since 1988 by the National Endowment for the Humanities to tag all of the world's literature, was another heavy user of SGML. In the last five years, the initiative has developed a compact tag set to foster more use. (Burnard and Sperberg-McQueen, 1995.) C. M. Sperberg-McQueen, a primary force behind the initiative, has become a pillar of the XML community.

XML AND E-COMMERCE

XML was built for applications that

- require the Web client to mediate between two or more heterogeneous databases
- attempt to distribute a significant proportion of the processing load from Web server to Web client
- require the Web client to present different views of the same data to different users

- use intelligent Web agents to tailor information discovery to the needs of individual users.

Each is relevant to E-commerce. As buyers compare products and prices from virtual catalogs (i.e., databases) maintained by a variety of sellers, process this information to determine the best match (usually in their own machines rather than distant servers), negotiate transactions, and track delivery and payments, they are using all the application types described above.

But to understand the potential effect of XML on E-commerce, it helps to look at consumer-to-business and business-to-business transactions separately.

These days, most consumer-to-business commerce requires the full-time attention of the consumer and the electronic attention of the business. Such trade is often little more than an advanced version of catalog shopping—only with a much-larger catalog and some ability to engage in long-dormant pricing behavior (e.g., auctioning off standard manufactured items). XML may permit software to scour the Web looking for purchasing opportunities that are specifically coded as offerings. XML pages with standardized tags, such as <Price> or <ModelNumber> could allow the search for and perhaps even the negotiation of best matches between buyers and sellers, presenting the buyer with a set of options for final selection and approval. Clothes, for instance, might be described in terms of a data set so complete (e.g., fabric, piece sizes, color) that a customer could simulate its appearance on a range of body types. Travel arrangements could be automatically calculated by mixing and matching the arrival and departure times of various segments. Much of the processing would move from the seller's server to the buyer's client machine, while the server would contain product information in a format most convenient for the seller. A third party could rate and otherwise compare varying offerings by their parameters (e.g., what colleges offer) and their performance (e.g., medical outcomes). Indeed, all that is required to justify XML is the need to describe in standard terms something that may inform or lead to a purchase.

The case for XML in business-to-business transactions is a good deal more straightforward, inasmuch as they are already becoming (1) completely automated processes and (2) are backed by standards (ANSI X12 in the United States and the United Nations' EDI for

Administration, Commerce, and Transport [EDIFACT] internationally, i.e., in Europe). X12 and EDIFACT specify digital formats used to encode key business documents, such as invoices, bills of lading, and payment transfers.

Business-to-business E-commerce is more complicated than business-to-consumer transactions. It is iterative and requires the participation of different people within sellers' and buyers' organizations, with each person contributing part of the transaction. There are two different types of business-to-business transactions: repeat purchases within a long-term relationship and one-time purchases. In the former, buyer and seller negotiate product attributes, prices, and terms of purchase, after which authorized buyer representatives (and in some cases sellers themselves) can trigger purchases of individual items. In the latter, the buyer usually specifies and sends requirements to several potential suppliers. After several of these submit bids by a specified date, the purchaser decides which bid to accept. One or more rounds of negotiations with one or more potential suppliers precede selection. In both transactions, once a supplier is chosen, goods or services are ordered, and sometimes partial payments are made before or during production. There are specific documents that must be exchanged between buyer and seller before goods and services are accepted and final payment is made.

EDI, in its current incarnation, has been pushed by large organizations, which want to decrease their purchasing costs and have the clout to make the smaller trading partners use EDI. But such EDI has severe limitations. First, its use of specific message formats imposes a strict structure on the transaction. Second, complex person-to-person arrangements must often occur before two business units can reliably use EDI. Third, it is expensive because it usually involves proprietary software and proprietary Value Added Networks (VANs) to translate messages among various EDI software packages and provide electronic mailbox hosting services for trading partners. Although Web-based X12 applications are being developed, these applications do not remove EDI's most important limitations.

XML would do away with today's EDI's limitation on the content of communications between buyers and sellers, as well as with the expense of VANs—and thereby boost E-commerce. It could allow any two buyers and sellers anywhere to communicate directly, using their own formats for documents and a common set of content tags, all supported by commercial software and without the need for intermediaries. To preserve existing investments in X12 data, its message formats and tags could be included in XML-based EDI applications. But XML would also allow buyers and sellers to do things now impossible with today's EDI, e.g., to include human interaction within the E-commerce transaction stream, as different people are presented with Web-based forms for inputs and approvals within their organizational units or functions.

WHAT THE WORDS MEAN

But first XML must cope with the well-understood fact that the specification cannot alone ensure interoperability. XML's "body of knowledge" must include detailed syntax and vocabularies for communities of users—and the definition of communities must partition the universe of users cleanly enough so that there is little ambiguity among users over which language to use in conducting which business.

Thus, standard DTDs and vocabularies must be available to users via some sort of repository. High-level and general repositories could be managed by standards organizations; industry-specific repositories could be managed by industry groups, and more specialized repositories could be maintained by groups of partners or within individual companies. Several standards, addressed to the needs of individual communities, have already been published through the W3C, including the Mathematical Markup Language, the Chemical Markup Language, and the Astronomical Markup Language.

But many more groups are developing DTDs, suggesting that XML may be a victim of its own early popularity. The old saw that the wonderful thing about standards is how much choice one has in them is, at this juncture, less than completely amusing. Take the following examples:

- The Open Trading Protocol is a consortium of banking, payment, and technology companies specifying information requirements for payment, receipts, delivery, and customer support.
- The Open Buying on the Internet initiative, launched by American Express, Ford Motor, Office Depot, and others is

automating large-scale corporate procurement of office and maintenance supplies.

- RosettaNet is a PC industry initiative, managed by a board of 34 chief executive officers and chief information officers of major information technology users and vendors, which defines how to exchange PC product catalogs and transactions among manufacturers, distributors, and resellers. RosettaNet participated in a pilot project with CommerceNet (a consortium of several hundred information technology companies) on catalog interoperability because the project included laptop computers.
- Under the rubric of the Information and Content Exchange, CNET (part of the News Corp), Vignette, and other information content providers are developing ways to create and manage networked relationships, such as syndicated publishing networks, Web superstores, and on-line reseller channels.
- The Open Financial Exchange, proposed by CheckFree, Intuit, and Microsoft, supports banking, bill payment, investment, and financial planning activities by consumers.
- A consortium of 40 companies, spearheaded by software vendor Ariba Technologies, has developed Commerce XML (cXML) to standardize catalog content and purchasing data exchange.
- Microsoft has its BizTalk initiative.
- In June 1999, J. P. Morgan and PricewaterhouseCoopers LLP announced the Financial Products ML, designed to address the needs of the financial derivatives community.

XML may be standardized for commerce if combined with X12. There, too, several groups compete with the others in that they take different approaches, yet all claim to cooperate with each other. CommerceNet's framework for open Internet commerce, eCo System, was originally (1996) based on CORBA and later (1997) recast on an XML foundation (thanks in large part to the support of the big software companies). This framework promulgates a set of Business Interface Definitions (BIDs), which, when posted on the Web, tells potential trading partners what on-line services a company offers and what documents to use when invoking them. Its Common Business Library, an extensible public collection of generic BIDs and document templates, includes XML message templates for the basic business forms used in X12 transactions. The Defense Information Systems Agency is funding more work into interfaces between XML and X12.

The U.S. XML/EDI working group was established in July 1997 (see XML/EDI Group, no date) with W3C's infrastructure support but with no explicit endorsement (this requires a formal working group recommendation to be submitted to a vote of the membership and then approved by the director). An international XML/EDI Group, housed by the Graphic Communications Association Research Institute (Alexandria, Virginia), is looking to create "a new powerful paradigm, different from XML or EDI" by "first implementing EDI dictionaries and extending our vocabulary via on-line repositories to include our business language, rules and objects." (Graphic Communications Association, 1999.)

Europeans have their own XML/EDI Pilot Project, under the European Center for Standardization/Information Society Standardization System (CEN/ISSS). They seek to "explore how XML can be used to provide an interface between existing EDI applications and the next generation of XML-aware applications" and study how XSL could help present EDI messages to people in ways that account for variations in their linguistic and cultural background."[9] It also comments on how the W3C's work on XML and EDIFACT can be used with "the multilingual and mixed trading practices found in Europe." (CEN/ISSS, 1998b.) Europe's work builds on other XML-EDI work, such as EuroStat and the Norwegian government projects on the interchange of statistical data, CEN TC2251 for health care informatics, TIEKE in Finland on transport-related messaging, EDIFRANCE on E forms, and UK/CEDIS on Simple EDI. The project's success factors include the quality of the XML DTDs it created, the acceptability of the software tools to end users, and the acceptability of XML as an alternative to today's EDI. (CEN/ISSS, 1998a.) The project published its preliminary findings in October 1998. Europeans worry that American efforts fail to refer to the relationship between X12 and EDIFACT—a poor way to promote globalization of commerce, which is a stated goal of many XML-related E-commerce efforts. (CEN/ISSS, 1998e.) The European Electronic Messaging

[9]Preceding quotes from CEN/ISSS (1998d).

Association EDI Working Group has proposed that the UN create and manage a repository of XML tags based on EDIFACT. (Raman, 1998.)

MANAGING PROLIFERATION

One approach to the problem of standards proliferation is the creation of ontologies (a concept from the study of the nature of knowledge), each of which codifies the concepts meaningful to a community. Thus, everyone would have a common understanding on which to build vocabularies. Ontology.Org and CommerceNet (Glushko et al., 1999) are working to create a set of business-related ontologies, such as various aspects of payments and business processes.

Another reaction has been the formation of consortia to develop and maintain a registry of vocabularies. OASIS is composed of vendors and consumers assembled to work on interoperability shortfalls between products or among software suites. The focus is on horizontal application products, such as XML table models or conformance suites. They are moving into registries, in what may be some competition with Microsoft's Biztalk initiative. As of mid-1999, the two efforts had become at least somewhat harmonized.[10] OASIS is tied into CommerceNet in that its Registry and Repository Technical Committee is (as of mid-1999) chaired by one of its employees.

XML AS A STANDARDS ABSORBENT

One sign of the hopes being invested in XML has been its ability to encompass other standards (e.g., SGML). Supporters of many other standards have hopped on the XML bandwagon by converting their vocabulary into tag sets, quietly chucking earlier vehicles. Many such standards, however, had yet to achieve much lift.

The W3C's Platform for Internet Content Selection (PICS), for instance, predates XML. It is a structured set of Web references and metadata tags through which Web sites could attach ratings (e.g., for movies) provided either by the site's owner or through an external

---

[10]Microsoft is a member of OASIS, but membership in a consortium has never been a bar to advocating an alternative standard. Although Microsoft is a member of the Object Management Group, it continues to tout its Common Object Model (COM) in competition with the latter's CORBA.

rating service. When developed, the standard was expressed as a parentheses-denoted Multipurpose Internet Mail Extension (MIME), type (an IETF standard designed to reformat 8-bit content into 6-bit legal characters used for Internet E-mail) and as META tags in HTML. Once XML was developed, however, PICS could be denoted as markup tag, and so parentheses were replaced by angle brackets and XML tags. But the words were the same. In time, the RDF (resource description framework) grammar will replace the PICS grammar, but, again, the words will remain. The world of digital libraries, as noted in Appendix C, provides further examples.[11]

Of note is HL7,[12] a standard way to specify and format messages to exchange, manage, and integrate data for clinical patient care (notably via admissions, discharge, and transfer systems). Although the standard has ways to describe the medical care given (i.e., what all the billing is about), it was not meant, at least originally, for doctor-to-doctor communications but for medical E-commerce. The standard appears to be well-established (the parent body, also called HL7, had 1,700 members in 1998), but the standard is not meant for casual use: Two parties who agree to implement the standard must write an auxiliary specification that specifies event triggers, messages, and optional fields used and omitted (so as to trim the broad list of data elements otherwise required). As with many heavyweight standards, HL7 is more suited for interoperability within an enterprise than among enterprises. (Lincoln et al., 1999.) Since starting in early 1987, HL7 has shifted from OSI to the now-ubiquitous TCP/IP. Moving it further to XML may represent a larger change because HL7, although transport-independent from its inception, was developed to encode messages according to strict rules.[13] Developing a DTD for HL7 and then extracting HL7's semantics apart from its syntax would be major changes that would have to be carefully engineered to ensure that the structural information in the current speci-

[11]For instance, ten years ago NIH adopted ASN.1 for PubMed classification. Having mooted CORBA, NIH is shifting to XML.

[12]The *7* in *HL7* refers to the seventh or application layer of the OSI model. Like OSI, HL7's developers wish to bracket the standard with reference models and usage profiles.

[13]As of 1998, developers looking toward HL7 version 3 (version 2.3.1 became an official ANSI standard in May 1999) were trying to put it over an object-oriented methodology. (See Hentenryck, 1998.)

fication is not lost in the new XML rendering (even as the overall HL7 message envelope persists).

HOW XML MAY FAIL

The most obvious way that XML may fail is that the promise of interoperability may be lost in the welter of competing semantic standards that use the XML syntax. But there are other ways to fail.

Other Standards for E-Commerce May Arise

Some proposed standards for Web commerce are incompatible with XML. A UN group is promoting Object-Oriented EDI (OO-edi).[14] OO-edi comprises two views: (1) a Business Operational View (BOV), which defines parties to the exchange, their roles, business processes, agreements, and data, and (2) a Functional Service View (FSV), concerned with implementation details, such as the syntax and method used, communication protocols, and application interfaces. The Universal Modeling Language was then selected for business process and information modeling. Although the group favors BOV, it avers that XML can be used with one of many types of FSV implementation. However, XML's use within an OO-edi environment would require a tricky data mapping to business objects, whereas pure OO-edi does not require it. (Harbinger Corp., 1999.) The UN group has not endorsed the XML/EDI Group promotion of XML as the FSV solution for OO-edi. (Webber and Naujok, 1998.) (The complexity of this paragraph provides a good hint about the standard's prospects.)

Business system interoperation (BSI) is an approach to EDI that uses BSI servers at each end of an E-commerce transaction for encoding and decoding. A perhaps fatal limitation of this method is that it requires exchanges of updates between trading partners every time one of them makes a change to its internal process or software. A project on BSI in the reinsurance industry is being supported under Europe's ESPRIT IV and undertaken by the Distributed European System Interoperability for Reinsurance (DESIRE) consortium.

---

[14]The Techniques and Methodologies Work Group (TMWG), charted by the United Nations Centre for the Facilitation of Procedures and Practices for Administration, Commerce and Transport (CEFACT).

Although CEN/ISSS initially supported BSI (see CEN/ISSS, 1998c) it formally withdrew from the project in mid-1998.

Electronic Data Markup Language (EDML) is a metadata coding system for use in defining the NAME component of the META construct in HTML. According to its creators, it is not intended as a competitor to XML but can be used as a stand-alone (Galbraith and Galbraith, 1998)—but if it works, XML may not be needed for E-commerce applications.

Such approaches are, at worst, distractions. The UN effort is clearly the work of structuralists who believe that a rigorous descriptive architecture of any realm must precede (or substitute for) its semantics.

Too Much Capital May Have Been Sunk into Today's X12- and EDIFACT-Based EDI

Companies that use EDI now have large investments in EDI software and may be reluctant to throw it all away. Major EDI service suppliers, like GEIS (General Electric Information Services), are developing Web-based EDI applications, which might prolong EDI's life. Since a transition from X12 or EDIFACT to XML requires some form of translation, at least for legacy systems, it is not clear that moving into XML-based commerce will make economic sense in many cases. To succeed, XML product suppliers will have to provide flexible and scalable interfaces with a variety of legacy business systems—an untested capability. Indeed, reducing the cost of EDI may not be in everyone's interest. Large firms may look at the cost as a way of testing the seriousness of a vendor's commitment, while vendors who have made the requisite investment can regard such costs as a barrier to new entrants. Finally, but by no means decisively, XML-based transactions will also require somewhat more bandwidth—one estimate is roughly 15 percent (EPIFOCAL, no date)—than traditional EDI transactions.

XML Is Still Too Complex

Because XML is not new, but a skinny version of SGML, it may not reduce the complexity of SGML enough. (Cover, 1998.) XML is itself complex, and many XML applications proposed include DTDs, themselves quite complex. XML has to feel right to the average HTML coder before it attains the ubiquity to replace HTML.

It May Get Caught in the Browser Wars

If HTML's history repeats itself, XML may suffer from having different browser makers include various nonstandard features. According to the Web Standards Project, an international coalition of Web developers and Web experts, Internet Explorer 5.0 does not fully implement the XHTML 4.0 standards that Microsoft helped develop. While some standard features are missing, others are implemented in a way that would make them incompatible with other standard-complying authoring tools. (Olsen, 1999; Bray, 1999.) Since Netscape announced that Mozilla will be fully compatible with the XML standards, a repeat of the "browser wars" may be in the offing.

It May Get Caught in the Java Wars

Combining XML-marked-up data with cross-platform software, such as Java, allows the formation of movable objects. XML is platform-independent data, while Java is platform-independent software. Sun's Director of Java Software, Jonathan Schwartz, maintains that XML, together with Java, can support the requirements for reuse of information across arbitrary and idiosyncratic computer systems and display devices. (Alshuler, 1999.) The combination would also result in acceptable implementations of object-oriented EDI.

So why is Microsoft embracing XML so hard in its Biztalk effort—which combines an active registry program with vertical marketing of Microsoft products into the E-commerce sector and efforts to make future browsers XML-aware? Even though there is no reason that Java code cannot work with XML-formatted documents, an applet-centered world and a document-centered world pull people in different directions.

In an applet-centered world, the server provides the data and the applet to manipulate it; the data need not be formatted in any fashion that outsiders have to agree to. Why? The definition and treatment of the markup come from the same institution that produces the applet. It suffices only that the applet recognizes what the tags mean; users do not have to. The wide use of the XML grammar can make applets easier to write because the tools to manipulate marked-up text will be widely available, but the words need not be standardized.

In a document-centered world, the tags would have standard meanings. That being so, off-the-shelf software can be built to recognize the denotations and connotations of the tags to manipulate the document. Applets are no longer as necessary because the manipulation capability can be built into the browser or an add-on. Thus, Microsoft's approach requires XML to push beyond grammar to words; Sun's approach exploits XML for the regularities in the grammar.

Sellers May Not Like Friction-Free Capitalism

Not every seller, after all, wants to be compared on the basis of a particular attribute to the exclusion of other attributes (e.g., revealing price but not customer support and thereby encouraging commoditization of the pricing structure). Nor do all sellers want to allow their sites to be searched by bots, thereby losing the ability to present their terms to human decisionmakers. With current technology, some sellers limit access to their sites for nonhuman visitors. When implementing their catalogs in XML, sellers might adopt nonstandard tags or might design their sites in a way that provides the information they choose to provide, regardless of the information requested, e.g., information on product or service bundles only. This is not necessarily a bad idea. Depending on the seller's brand and market power, it may be in a position to demand and get different trading terms than less successful competitors. XML provides a possibility of a level economic playing field in which consumers would benefit; it does not necessarily create conditions under which sellers will want to play.

Trust, Not Standards, May Be the Problem

Here too, XML alone may not suffice until and unless issues that relate to the social aspects of business are put to bed (see Appendix D's discussion of security and payments). One such issue is trust. Will every buyer that contracts for a purchase have the funds to pay for it? Will sellers deliver the promised goods on schedule and at expected quality levels? It is always risky for new buyers and sellers to transact business until they build a record of fulfilled transactions and trust. Part of the "value added" that such intermediaries as General Electric Information Services provide is the screening of buyers and sellers, increasing comfort levels for both parties. While a global market is a theoretical nicety, relying on the kindness (or probity) of strangers is still a lot to ask.

CONCLUSIONS

XML, if it works, may very well be the heart of tomorrow's Web because documents structured in a standard can be understood and thereby manipulated by stupid but fast and cheap machines rather than intelligent but slow and expensive humans. But despite the enthusiasm with which XML is being offered to, and, accepted by the world, the hard work lies ahead. Whether the XML standards processes can result in commonly defined terms within (and, perhaps more importantly, across) the disparate communities of commerce is yet to be determined.

Appendix C

KNOWLEDGE ORGANIZATION AND DIGITAL LIBRARIES

The dream yet lives of a universal knowledge base that the Web, uniquely, has the power to bring together. Achieving this dream has three parts: universal access to knowledge, protocols to exploit such access, and enough organization to find and categorize what is out there. Libraries have been the traditional source of knowledge, and standards activities to help their work are still going on. But changing the artifacts by which knowledge is conveyed from books to electronic bits also changes how institutions manage knowledge and thus the standards that would best help them do so.

THE CRISIS IN ACADEMIC PUBLISHING

One way to measure the success of standards for E-commerce is by how much money changes hands; the more money, the more success. Ironically, as the drive to liberate knowledge from its artifacts gathers steam, the reverse may better measure the success of standards in facilitating knowledge organization: the less money, the more success.

For centuries, books were the only practical method for transferring large amounts of information. Publishers and bookstores arose to sell these artifacts, and libraries arose to store them. In today's era, the artificiality of this arrangement is becoming clear. Words differ from books. Their pricing, the property rights inherent in their expression, and the challenges of their distribution follow from the physical form they take. Words liberated from paper can be analyzed and manipulated in altogether new ways. Yet, the institutions that traffic in the written word (including journals) and the market models consistent with converting information to artifacts are not going gently. Indeed, the earliest big success in electronic commerce, Amazon.com, is in the business of selling bits in the form of atoms.[1] If there is another viable market model for information, it will have to have its own standards. Information is not yet free to be free.

In the oft-quoted words of Samuel Johnson, "No man but a blockhead ever wrote, except for money." Implicit in the notion of intellectual property rights in information is that authors must be compensated with money lest their creative incentive wither. It is nearly costless to copy information electronically, but without a market model or technology that makes such copying difficult (or irrelevant), there is likely to be little incentive to liberate the content of information from a form (i.e., books) that frustrates widespread duplication.[2]

Academic publishing does not fit this model. Authors are rarely compensated for their contributions and, in many cases, must pay journals to print their submissions. Scientists and other academics publish to document their work, foster the accretion of knowledge, and win the acceptance of or prestige from their colleagues.

Nevertheless, the cost of academic literature has been rising much faster than the rate of inflation,[3] particularly in the last five years, reaching roughly $2.5 billion a year in the United States alone. Hence, a crisis has arisen among university libraries. They have either had to drop their subscriptions or find other funds with which to buy them (often at the expense of monographs). (For instance, see Faculty Taskforce, 1998). Many of the journals whose subscriptions

---

[1]In retrospect, the Web is an obviously good way to conduct commerce in which the number of different products on offer is very high—millions in the case of books. In further retrospect, this is why EBay works: The number of individual offerings is immense.

[2]The music industry appears to be in the lead here in terms of watermarking files so that they may be traced back to their original buyer. See the Secure Digital Music Initiative in Appendix D. Text files, because they are relatively small and every bit counts, do not lend themselves to watermarking so well.

[3]Since 1986, the 121 members of the Association of Research Libraries have spent 124 percent more on serials to purchase 7 percent fewer titles. (Renfro, 1997.)

have been dropped have had to stop publishing.[4] The high cost of subscriptions prevents their circulating through much of the third world, leaving behind overseas colleagues whose potential contributions to the corpus of knowledge are thereby later and less pertinent.

In Spring 1999, Harold Varmus, who heads NIH, proposed that the distribution of biomedical research be shifted from physical to electronic form, as E-Biomed (see NIH, 1999). NIH, which invests $16 billion a year to advance the state of knowledge, had an understandable interest in seeing the use of its research unfettered by irrelevant commercial considerations.

NIH's PubMed already indexes and abstracts nearly all of the world's medical research as it is, and publishes full-text versions of a large percentage, but by no means all, of its corpus.[5] In one field, physics, preprints have been published electronically for years.[6] The NIH proposal envisions two types of records: (1) those that have been peer-reviewed either by extant editorial boards or by boards established by E-Biomed's governing board or (2) unreviewed papers judged only for appropriateness by two relevant individuals. David Shulenburger, Provost of the University of Kansas, has proposed a more radical solution: establishing a National Electronic Article Repository, backed by a federal law that would mandate that anyone receiving federal money for *any* research submit an electronic copy of everything to be published in a journal within 90 days of its being printed. In November 1998, the Kansas Board of Regents made such submission an official policy among its university faculty.

Although electronic publishing would reduce the costs of scientific publishing for those who could live without hard copy, the competition may also temper hard-copy prices for everyone else. Some publishers have gone so far as to define electronically posted documents

---

[4]See, as a general treatment of this subject, Darnton (1999).

[5]Many publishers provide citation information electronically to PubMed because they feel a PubMed citation provides the visibility they want. It helps that PubMed is not a potential commercial competitor.

[6]The archive, started in 1991, has been supported by the National Science Foundation since 1995 on a contract extending through 2000. As of April 1999, 100,000 papers have been archived, in fields that now include mathematics and computer science. The archive has not allowed indiscriminate downloading by bots since March 1994. (See Ginsparg, 1996.)

as having been published "elsewhere" and therefore exclude them. Nevertheless, the number of electronically published journals has increased; by late 1997, the number of refereed journals had exceeded a thousand (as against the 14,000 in hard copy). One service, JSTOR, provides electronic access to 117 journals (as of June 1999) at an annual subscription cost of $5,000 a year (less for smaller libraries), but it did so through the initial support of the Mellon Foundation.

Drawbacks to a hasty transition from print to electronic storage have been voiced.[7] The profusion of electronic material these days has given the entire enterprise a bit of a fly-by-night scent—but nothing that the right kind of institutional backing cannot fix. The crux of the objections relates to peer review.[8] The NIH proposal calls explicitly for peer review on one part of the archive[9]; one advantage of electronic format is that it makes postpublishing review and commentary a good deal easier to access.[10] Cross-referencing is generally easier in electronic form; so is indexing by topic or even by included text (*if* the material is rendered in searchable text form and not as an image). Another great advantage of electronic over hard-copy publishing is that the former can include videos, simulation, or models within the text.

So, where is the standards angle? Existing Net and Web standards (plus search engines powerful enough to sweep the Net regularly) are sufficient to permit the transition from atoms to bits in holding the ever-growing corpus of scientific and academic literature. What standards do, however, is accentuate the potential advantages of electronic publishing by making it easy to find and organize elec-

---

[7]Incidentally, the oft-cited objection that paper is easier to read is irrelevant; indeed, machines exist (e.g., the Xerox Document Binder 120) that not only print but bind electronic material into a book. Research libraries may not necessarily benefit from electronic journals, if they are directly accessed by users leaving libraries disintermediated from the process.

[8]See, for instance, Pear (1999).

[9]Objections were raised that the unreviewed part of E-Biomed would be a government-supported repository for junk science. Doctors may read the material and call for harmful treatments as a result.

[10]Although there are, of course, no tags from previously printed material to future commentary, search engines could be used to find such commentary as long as the references to the material are consistent.

tronic material. And that takes us to the world of libraries, whose habits, structures, and standards are not necessarily attuned to the Web.

THE WAY OF THE LIBRARIAN

Librarians have been the traditional custodians of knowledge, or at least its artifacts, and their energetic endeavors on behalf of standards are still the leading source of activity in this field. Within the United States, the National Information Standards Organization (NISO) and its standards (e.g., the Dewey Decimal classification system) dominate the world of libraries. Although not strictly a national library (such as Canada and France have), the Library of Congress is the de facto institutional leader of U.S. libraries.

One area in which interoperability in the physical world matters is interlibrary loans. Such activity is supported by a standard access format (Machine-Readable Cataloging [MARC], now under the aegis of the Library of Congress) and a common carrier, the On-Line Computer Library Center (OCLC). Originally established as a simple clearinghouse, the OCLC grew into an X.25-based network that provides a shared catalog service (6 million records from 2,500 collections) and interlibrary loans. (Horny, 1984.) MARC and OCLC both started in the mid-1960s.

Among the more prominent library standards is NISO's Z39.50 query-and-retrieval system. Z39.50 started in 1980 with the Library of Congress's Linked Systems Project (joined by OCLC in 1984). The project's initial purpose was to link catalogs of the Library of Congress, the Research Libraries Group, and the Western Library Network.[11] Early in the project, participants developed a protocol for intersite search and retrieval of records that in 1988 became Z39.50. In a Z39.50 session, the asking system requests data conforming to certain criteria; the responding system describes the size and composition of the responses, and the asking system indicates whether it wants to see all, some, or none of them. While that project was under way ISO put out its standard, Search and Retrieval (ISO 10162 and 10163). Although Z39.50 converged with ISO (version 2

---

[11]See Dempsey (1992), especially Chapters 4 and 6. Progress was slow, because of coordination problems and the initial decision to build upon the OSI suite.

became fully compatible), it retains services the ISO standard does not support.[12] The most recent (1995) version is a compatible superset of the 1992 version.

The Global Information Locator Service (GILS) (né the Government Information Locator Service) is one Z39.50 implementation, developed to maintain voluminous data on global warming (so, the U.S. Geological Survey administers it) and now in general use within the federal government, but rarely outside it. GILS is a standard both for compliant servers and for the records that they hold. GILS-aware clients query these servers to discover what information they hold. Several commercial-server software implementations now couple GILS standards with AltaVista search engines.

Uniform protocols help bridge differences among catalog systems, in particular how loosely or tightly book requests are mapped (e.g., keyword = automobiles and author = Cole) into lists returned. The fate of standard search methods depends on competition with proprietary search methods. Many important sites have committed themselves to standards, and Z39.50 has been adapted for the Web. Success may depend more on the perceived advantage of the client-server model or the need to link disparate systems with consistent communication interfaces than on any specific format; Z39.50, for instance, was delayed by being linked to OSI.

Standard references, query systems, and document formats are only of modest help, though, when seeking the *contents* of a requested document. This quest may instead require software that uses a formatted inquiry to look up pieces of information best suited to the user's interest. It requires a common format and a universe of uniform references with uniformly represented contents—an index or a fact-filled summary.

The Wide-Area Information Server (WAIS) returns document names and content in response to queries.[13] WAIS extends Z39.50 (1988 not 1992) by using a generic language to query text-oriented databases

---

[12]Z39.50 allows the target system to ask users to authenticate themselves and allows sending interim status reports during long searches.

[13]The WAIS concept came from Brewster Kahle, who was at Connection Machines and had help from Apple, Dow Jones, and KPMG, an accounting firm. WAIS was a good fit for Connection Machines, whose computers were efficient at simple but voluminous text comparisons.

often maintained in proprietary formats (by, e.g., Dow Jones News Retrieval, CompuServe, Dialog, and Mead Data). WAIS indexes every keyword in a document and divides the indices among servers organized by a top-level directory. Searching is a two-level process. A WAIS server responds to a query by applying a requested word set to a full-text index of a database, then ranks matches using relevance feedback (in which documents are selected if similar to others previously marked as relevant). (See Schwartz et al., 1992.) A WAIS-generated document index can be as large as the requested document itself.

WAIS servers tend to be ambitious because text data have structures more complex (e.g., base text, headers, footnotes) than, say, data tables. Text can generally be retrieved by reference (e.g., title), association (e.g., hyperlink or index or both), or content and criteria. Once search engines leave the world of literal and thus deterministic matching,[14] the task of matching criteria can be tricky—for example, list all articles written in 1990 that predicted the breakup of Yugoslavia. Standardizing search algorithms to return predictable article sets would be extremely difficult and would require either natural language understanding or embedded tagging (e.g., as in XML) that links documents to what they cover. Most search engines on the Web are patterned after the WAIS method of searching, indexing, retrieval, and relevance ranking (whose 0–1000 rankings can be seen in some search engines' results). WAIS now manifests itself as the code upon which proprietary extensions have been developed—and not so much a standard per se.

New methods—prototype standards as it were—for document retrieval are still being funded. The NSF, NASA, and DARPA have teamed to foster the Digital Library Initiative. Its first phase brought together the University of California at Berkeley, the University of California at Santa Barbara, Carnegie-Mellon University, the University of Illinois Champaign-Urbana, the University of Michigan, and Stanford University; the last will handle interoperability issues—an area to receive greater emphasis in the second phase. The current version of SDLIP (Stanford Digital Library, no date) describes how information clients can use a Library Service Proxy to query the contents of an external information source. The protocol, both CORBA-

[14]WAIS, for its part, does not automatically suggest synonyms for search terms.

and XML-ready, covers the transaction: how to specify quality of service, payment, whether or not to forward a list of all or just part of the documents, how long the server should remember the client's request, and so on. The SDLIP document often refers to usability on a thin client (e.g., one with little memory). This requirement lends SDLIP much of its complexity. It is redolent of the days when complex library catalogs were accessed through limited terminals. Today, even palm-sized devices have more than enough memory to handle text files much larger than a reader may wish to peruse at one time.

Exactly what problem is a digital library standard supposed to solve? Many library standards have been driven by the interoperability requirements of interlibrary loans (which, at most, accounted for only one of 30 books borrowed). Outside that setting, the only purpose of standardizing files and retrieval methods was for software and training portability—a far less urgent requirement. In cyberspace, however, books and information about books are simply bits. At a trivial level, there is no interlibrary loan[15]; at a somewhat less trivial level, there is no essential difference between abstracts, articles, and books themselves. Furthermore, both the software and hardware exist for near-instantaneous full-text searches of enormous databases (e.g., the roughly 320 million pages on line as of mid-1999). Nor is storage a problem: If, by one estimate, the Library of Congress holds 25 terabytes of textual data, a mere $160,000 worth of rotating instantly accessible hard disk capacity (or 5,000 DVDs) suffices to hold it all.[16]

Document access has become not a technical but a social and institutional problem. One gigantic server could very well hold the

---

[15]Indeed, there are no loans, as such, unless copies can be engineered to self-destruct. This would require either that documents come equipped with the appropriate operating instructions (which current documents lack entirely, unless laced with macros) or that users be persuaded to access material through a device that destroys such material in time. Digital video express (DIVX), a form of digital video disk (DVD), uses a similar logic but has faced fatal market resistance. It may be hard to persuade users to read material through a hostile device if it can otherwise be read with a more benign browser. Time will tell if the newest generation of hand-held electronic book readers will (1) sell well and (2) have adequate provisions for preventing a book from being copied over to the Web.

[16]That is, 600 hard drives of 40 gigabytes each at $270 per (as advertised by Office Depot and other vendors in December 1999).

world's entire library collection (even as other servers each hold subsets thereof). If so, whatever access method it adopts is the de facto standard for retrieval. Or there could be one recognized search engine to which various libraries would periodically submit a searchable index (or from which one periodic bot request would be issued).[17] Again, the standard would be whatever the search engine decides to adopt. Granted, neither model permits libraries to collect money from having their catalogs searched (and SDLIP explicitly includes provisions to enable money transfers), and neither necessarily provides libraries much information on which documents have been accessed. But whether libraries, especially those owned or supported by governments, should be charging for such services is a question that precedes standards. Need the documents "held" by a library be immediately accessed only through the library?

THE SEARCH FOR STRUCTURE

The printed word stands by itself; scholarship is what associates bodies of information with each other. Hypertext is a world of explicit links; compilations and indexing are others. Neither such linkage is going to happen for free, and the task of standardizing the content of information rather than simply its container (whether electronic or otherwise) is a Herculean one.

Most browsers permit users to find documents that contain certain words (sometimes combined with site names). XML permits authors to insert descriptive metadata. The accompanying requirement, if XML is to help with retrieving knowledge, is a standard way to query XML-formatted material. Accordingly, in 1988, Microsoft submitted a position paper to the W3C that called for a query language (XQL) based on XSL (designed to convert markup conditionally into presentation), in contraposition to a more complex and essentially moribund AT&T proposal for a query language (XML-QL) based on SQL (which was designed to extract data from data tables). The requirement for a formal query language standard is stronger if queries are resolved by the server (which must then recognize the

---

[17]Mike Schwartz (University of Colorado, Boulder) developed Harvest, software that permits a Web site to index itself and ship the information on request to Web sites, presumably in lieu of entertaining their bots.

client-generated query) rather than the client. If clients retrieve documents and then query them, the query takes place on the client's machine: Understanding the markup is good enough. In that case, a standard query language is whatever the browser (or any other client software) says it is—as long as its claims correspond in some degree to the user's expectations.

So far, digital documents, notably those on the Web, have even less structure than books do. Under OCLC auspices, a February 1995 conclave developed the Dublin Core, a primary set of elements to characterize such documents (the MARC list was perceived as too heavyweight and syntax-dependent). The list of 15 (expanded from 13 in 1996) is straightforward (Weibel and Hakala, 1998) and, to the designers' credit, remains compact. Seven terms describe content: title, subject, description, source (from which the item may have been derived), relation (the item's relationship to the source), language, and coverage (where and when does the story, as such, take place). Four terms describe intellectual property: creator, publisher, contributor (e.g., an editor, translator, or illustrator), and rights (i.e., how the item may be used). Four describe instantiation: date, type (e.g., home page, novel, poem), format (which suggests the software required to read it), and identifier. The syntax to carry these semantics was left unspecified. Early implementations invoked the META tag of HTML, the interface definition language of CORBA, or the multipart type of MIME. Currently, XML looks like the carrier of choice. The library community, no strangers to the pleasures of good order, are trying to standardize the Dublin Core through the IETF (RFC 2413 and a May 25, 1999, draft to encode the Dublin Core in HTML), the NISO, and CEN.

The definition of legal data entries for each term remains open-ended. For instance, there is no specified list of subject categories. Thus, a problem remains that plagues Web searches: The word *mercury* may refer to a planet, an element, an automobile, a Greek deity, a space capsule. At least the Dublin Core lets authors categorize their material to differentiate the articles about *Mercury* from articles containing the word. And down the road, there may be standards by which Mercury-as-a-Greek-deity can be distinguished from mercury-as-an-element in a subject—perhaps through hierarchical decomposition (e.g., subject: planets, Mercury, etc.).

The Warwick Framework, born a year later, reflected a desire for a higher-level context for the Dublin Core. The idea is to define how the core and similar constructs can be combined with other metadata sets so as to retain their integrity for distinct audiences and separate realms of responsibility. (Lagoze, 1996.) The Warwick Framework specifies a set of containers for aggregating distinct packages of metadata (and other containers).

The W3C's Resource Definition Framework (RDF) of February 1999, a parallel recommendation, was also shaped by the Dublin Core community. RDF specifies a syntax by which various tag libraries (e.g., DTDs) can be referred to by acronyms so that documents with heterogeneous tag sets can (1) distinguish the source of each tag and thus its meaning and (2) distinguish among tags from two different sources that have the same name. Using RDF, a DTD could associate "dc" with a Web site for the "Dublin Core" and then use a tag "<dc:title . . .>" to mean "title" as defined with respect to the Dublin Core.[18] RDF also has ways to group values together (as sequenced sets, unsequenced sets, or sets for which only one element is applicable), to describe relationships among properties. Ironically, the paired standards of the Dublin Core and the Warwick Framework neatly bracket the minimalist-structuralist continuum of philosophies on how to construct standards.

Can knowledge ever be unified? Aristotle was said to be the last person who knew everything known to his civilization. As knowledge has grown, even knowing who knows everything in ever smaller subfields has grown impossible—as if such designations would be accepted in the first place. The Internet, as an institution, has stood for the principle of convergent organization that permits information to cohere magically as long as there is adherence to a few simple standards. But the task of developing ontologies through which knowledge can even begin to be classified is daunting and a deeply structuralist project. What few proposals exist for expressing ontologies, such as the Knowledge Interchange Format (KIF), or manipulating them, such as the Knowledge Query Markup Language (KQML), remain academic playthings. RDF may help, but only by

[18]For example, <Description about = "document" <DC:title> title </DC:title> </description> etc.>.

sorting out whose descriptions are being used ("is this description of a CD-ROM tagged by the vendor community, the payment community, the music community, the critic community, or what?").

TAGGING INFORMATION WITH PROPERTY RIGHTS

Which brings the discussion full circle to the community of those who do *not* want their information to be free: professional authors, publishers, and other rights holders. This is the world of Digital Object Identifiers (DOI), and the Interoperability of Data in E-commerce Systems (INDECS) effort.[19] Advocates of INDECS were unhappy with the Dublin Core, which they saw as biased toward text (as one would expect from librarians) and insufficiently sensitive to the many nuanced relationships among authors, creators, contributors, publishers, and so on (it is, for instance, utterly incapable of correctly rendering the particulars of even a modestly complex Hollywood deal). To INDECS people, the point is not so much to find things but to make sure that users understand the intellectual property rights of things that they find.[20] In developing standard metadata to protect their equities, they started with a generic metadata schema. It elaborates all possible relationships between content (which could be copyrighted material, performances of copyrighted material, reviews of performances of copyrighted materials, etc.) and then moving upstream to people (which would also entail many people, nonpeople, etc.) and deals (the many ways that people can hand over value for content). One sample construct is the relationship between a *work*, which is realized through an *expression*, embodied in a *manifestation*, and exemplified by an *item*.[21] And then it gets more complicated (e.g., in some instances, the intellectual property belongs to the instrument—such as a telescope—that

---

[19]An initial membership list for INDECS draws heavily from those—predominantly from Europe—who make money from intellectual property.

[20]Rights need not imply payment. One example is an author's right to be cited as such if material is placed in the public domain. In Europe, "moral rights" associated with intellectual property inhibit others from inappropriately changing the form or substance of such material without the creator's consent.

[21]This example comes from the Functional Requirements for Bibliographic Records (FRBR) model of the International Federation of Library Associations (IFLA). (See Bearman et al., 1999.)

captured it). As of mid-1999, the metadata to support INDECS has yet to be written.

A DOI is a Universal Resource Name (URN) designed to manage copyright and provide documents some persistence and uniqueness—a problem in a universe of dead links and duplicate URLs.[22] The initiative was conceived in 1996 and announced in 1997 (at the Frankfurt Book Fair); 400,000 identifications were logged the first year. A DOI identifier has two parts: the number of the assigning authority, whose allocation is administered by the International DOI foundation, and then a number as determined by the assigning authority. The identifier's metadata is a minimal kernel of elements under an umbrella data model derived from INDECS analysis. (Paskin, 1999.) So far, the DOI and URN are proposals[23] and concepts, respectively. DOIs, in turn, are to be resolved to URLs through their naming authorities. A draft proposal to do this, called the Naming Authority Pointer, has been developed through the IETF and is based on DNS (RFC 2168, experimental).

Like telephone numbers, Internet addresses, and Library of Congress designators, the DOI is a standard whose implementation requires some authority (whether unitary or hierarchically federated) to evaluate the application for a unique identifier and award one. This guarantor, the DOI Foundation was (as of mid-1999) still working out how it will be paid for its services. The Corporation for National Research Initiatives developed its Handle System starting in 1994 as a specific technical implementation—an open set of protocols, a namespace, and a reference implementation—of a uniform resource name of which DOI is a subset.

The biggest problem with all tag sets is getting people to use them. Advocates of the Dublin Core, by keeping their tag set short and simple, must be hoping that if the burden is light, compliance will

---

[22]The URN, an oft-mooted but not-yet-well-implemented concept, can cover any naming system (e.g., telephone numbers) not just digital objects. For instance, urn:doi:10.1000/123456789 would be a four-part DOI name in URN syntax (the "/" separates the naming authority from the object identifier).

[23]The proposed list includes genre (what kind of object?), identifier, title, type (is it a work, physical manifestation, digital manifestation, or performance?), the primary agent, and the role this agent plays. Academic Press has recently announced it would assign DOIs to every article carried in its IDEAL electronic journals.

follow. Nondisplayed tags are unlikely to arise without some systematic way to read and react to them. Widespread insertion of such tags is unlikely until XML is built into browsers, search engines, plugins, or Java programs that will know how to do useful and interesting things with such tags. By contrast, the INDECS community, should it succeed in actually generating tags, is more likely to be motivated by money (and will thus be applied, not by the creator, but by those who would assert the rights to the digital product). It is easy to foresee a day when publishers will be forced to tag materials (with INDECS tags?) on the advice of lawyers who argue that their rights would be alienated by failure to append precise legal language to their documents. The legal precedents that give wind to such arguments have yet to be set.

CONCLUSIONS

Many standards intended to govern the search for and thus the metacontent of intellectual property will respond to the needs associated with *free* information—at least until copyright owners of digital books (and bit fields of similarly archival quality) find a good market model for their wares.

To date, a growing share of all public information is free: government documents, informal writings from academic and quasi-academic (e.g., nonprofit) institutions, 'zines, texts written before 1914, and explanatory material from commercial sites. Metadata (e.g., catalogs of libraries and museums), when amassed, are also likely to be free. Formal academic literature (e.g., prepublication papers) may well join the pile.

Periodicals and newspapers could follow. The low cost of duplicating information—that is, its marginal cost—is falling below the cost of billing and administration required to collect money from it (this may argue for a micropayments device, but most such devices have died young, as the next appendix relates). Today's market model is trying (and often succeeding) to make a go of giving information away and making money from advertisement and the sale of user profiles based on their click streams.

But it is unclear whether general-market books and other material (e.g., imagery, music, and videos) will ever be free. And if they are not free, it remains to be determined whether there is a market model for selling them, much less lending them, as bits.[24]

That being so, digital libraries are likely to face the existential question of why they exist, and all standards that focus on the intellectual property rights inherent in digital data may very well be beside the point.

CODA

The noisy world of electronic commerce and the quiet world of digital libraries would appear to have nothing in common, but, in cyberspace, they are developing a point of tangency with the world of standards.

In the physical world, there are large businesses and small businesses. One factor that helps the small fry survive (albeit sometimes as franchises) is that they supply neighborhoods with services that people must travel to get—a cost that keeps them competitive against larger enterprises with economies of scale. In the virtual world, travel is costless, and thus virtual storefronts could, in theory, take over the entire market, leaving the small fry having less to do. In cyberspace, the small fry have to rely on specialization in niches where detailed expertise can be brought to bear. If there are to be many small businesses in cyberspace there must be a comparable number of niches.

So, how are the small fry to be found? In the physical world, the search space is small (e.g., the neighborhood may have only a few florists). In the virtual world, the search space is potentially huge, and finding any given niche requires some ordering among all of them (hint: it is more than looking up a name). Something closer to a hierarchical decomposition of a conceptual space may be needed.

This is exactly how books are found. And thus, down the road, both fields need some way to organize knowledge in some standard way—

[24]Even if there *is* a market model for selling information of rapidly depreciation value (e.g., investment gossip) simply because, by the time copies circulate, the information is too old to be useful.

in one case, a knowledge of commercial niches; in the other, intellectual niches.

Appendix D

PAYMENTS, PROPERTY, AND PRIVACY

The current struggles of payments, intellectual property, and privacy are further proof that standards live and die based on what—and whose—problems they solve. A standard that solves a problem that few care about is unlikely to garner critical mass. Size, per se, is not the issue. A little-used standard can survive within a cohesive group within which it faces little competition (e.g., scientific standards). But if the standard makes sense only within a broader arena (e.g., privacy conventions), it needs appropriate mass.

The most salient fact of E-commerce for this case study is that credit cards work (in the United States). They handle small transactions poorly, but, for reasons examined below, the market for small transactions is weak. Credit card sales may not be particularly anonymous (and thus not private), but most people do not seem to care enough. Perhaps they should. Or perhaps they really do—many polls find huge majorities who worry about the assault of technology on their privacy—but somehow have yet to find the right mechanism for assuaging their concerns. Are standards the right mechanism? How far can standards be used in lieu of explicit public policy?

In this realm, the standards battles that normally involve engineers and corporate representatives have two more disputants: the federal government's law enforcement and record-keeping bureaucracies, and activists who see standards issues as emblematic of a broader issue within the political and social agenda. Law enforcers have tended to favor strong intellectual property protections and weak privacy protections; activists, the reverse. So, many privacy standards controversies, such as those over encryption, have four, rather than only two (technologists and managers), classes of participants.

PAYMENTS AND PROPERTY

Payment mechanisms are social constructs whose existen...
on national cultures, institutions, regulations, and the accidents of economic history. The modern credit card was popularized by the Bank of America in the 1960s (perhaps the Graduate heard that his future was in "plastic" rather than "plastics") and is, well, everywhere. Japan, by contrast, is still a cash-based society: Law and custom discourage credit, and with low crime levels, people feel safe carrying great wealth in their pockets. Europeans have taken to smart cards (i.e., think phone cards that work on more than phones); similar technology flopped when experimented with on Manhattan's Upper West Side and the 1996 Olympiad in Atlanta.

Cyberspace could very well have developed a payment mechanism uniquely suited to its ethereal existence and free from the shackles of corporeality. But with the Internet born in the United States (which still accounts for more Web use than the rest of the world combined), the American way of doing things—plastic—had to be considered an early contender in the Web payments sweepstakes. Credit cards, though, were perceived to have three fundamental problems in cyberspace:

- Because of the anonymity of cyberspace, the potential for fraud was perceived to be much greater.
- They were inefficient for small transactions.
- Ironically, credit cards were not, themselves, as anonymous as cash, and transactions could, if amalgamated, create personal profiles in ways that violated one's privacy.

Secure Electronic Transactions (SET)

In response to the first problem, Visa International and Mastercard International established SET in 1996 to foster "the development of a single technical standard for safeguarding payment card purchases made over open networks." It was designed to

- establish industry standards to keep order and payment information confidential
- increase integrity for all transmitted data through encryption

- provide authentication that a cardholder is a legitimate user of a branded payment card account
- provide authentication that a merchant can accept branded payment card transactions through its relationship with an acquiring financial institution.

A SET transaction (see Gruman, 1998) is preceded by the following steps: (1) the customer opens a Mastercard or Visa bank account and (2) receives a digital certificate; (3) third-party merchants also receive certificates from the bank. In the transaction itself, (1) the customer places an order over the Web (among other means), (2) the customer's browser receives and confirms from the merchant's certificate that the merchant is valid and then (3) sends the order information to the merchant, who (4) verifies the customer's identity by checking the digital signature on the customer's certificate and (5), if satisfied, sends the order message along to the bank, which (6) verifies the merchant and the message and (7) digitally signs and sends authorization back to the merchant, who then fills the order.

Ostensibly, SET is a standard with reasonable scope and goals. But the origin of the system, as well as its costs, kept it from popularity. Technologists have never really embraced this approach, because it was not open. SET's complex structure resulted in additional overhead for the transaction, and key management remained difficult. Merchants objected as well: The implementation did more to protect the role of the middleman in the transaction than to ensure the smooth implementation of E-commerce. Criticism was also focused on the choice of certificate authority—the banks and credit-card companies—as the definer of the electronic wallet used to store materials safely on the host computer. Merchants quickly perceived that Visa and Mastercard were trying to have merchants pay the costs of keeping the two groups of credit-card companies in the role as middlemen.

SET's failure so far means that the credit-card companies are left holding the bag for fraudulent credit-card use (at least in the United States, which still accounts for the bulk of consumer E-commerce). The authentication mechanism of a physical signature is absent, and customers have no motive to trouble with an electronic signature (their liability, after all, remains limited to $50). Whatever security customers might receive knowing that the merchant is authentic is too modest to worry about. Consumers are also protected from fraudulent vendors by credit-card companies. Although desiring the security of SET, credit-card companies still profit from E-commerce, albeit somewhat less than if fraud were smaller. Activists have had little interest in SET and have instead been focusing on other areas of E-commerce—particularly, privacy matters.

Digital Cash and Micropayments

Because credit cards are poorly suited for small transactions, there *appeared* to be a demand for a payment mechanism that required far less overhead and provided the same anonymity as currency. Such a payment mechanism has long been assumed to be the digital version of cash, a form of currency that authenticates itself. In effect, the buyer would send the merchant a string of digits that signified the willingness of a third party to redeem this string with actual money. As such, the merchant only had to verify the third party—of which there would be a limited number—and not the buyer—of which there may be, ultimately, billions. Anonymity rests with the ability of the string to hide the trace of first issuer (i.e., who the third party gave the string to in exchange for real money) without eliminating the proof that the third party will redeem the string for money. Even the IETF has tried to accelerate the process.[1]

Such a scheme requires a standard representation so that it may be freely exchanged with others and easily manipulated by software. It also requires one or more third parties to issue cyber-scrip, and, in most cases, a firm that licenses the technology and the aforementioned third parties. Such a firm could easily profit handsomely—one good reason digital cash schemes have emerged not through an open standards process but from the evangelical work of entrepreneurs. Success might bring de facto standardization, giving way perhaps to de jure standardization of the technology.[2]

---

[1]In 1997, a proposed standard RFC was filed on the Open Trading Protocol (OTP) to encompass a variety of payment systems, e.g., Secure Electronic Transactions, Mondex, CyberCash, GigiCash, and Geldkarte. In 1998, the IETF formed a working group on the issue; in 1999, there was still no standard or any draft standard on the subject.

[2]As of mid-1999, some standardization work was ongoing to define a format for encoding microtransactions using markup language.

Success has been elusive. First Virtual, an early entrant, abandoned the business. Digicash, a David Chaum project (see Chaum, 1992), entered Chapter 11 in November 1998. Cybercash has dropped its micropayment business to concentrate on software. Not everyone has quit. Compaq (with its MilliCent software) is using Japanese government funding to investigate the Japanese market, and IBM is putting its toe into European waters. But neither is particularly optimistic and the U.S. market is dormant.[3]

What happened? E-commerce generally deals in two types of products: those that come on a truck and those that come through the wire. The latter may be divided into tickets (i.e., information that someone is eligible to receive a service, such as air travel) and information itself. Generally, anything that requires producer haulage or customer travel to consume is likely to cost enough to justify credit-card payment. The real micropayment market was thought to be small units of information, such as the permission to read an article. But people do not like to pay in drips and drabs; not for nothing have most Internet Service Providers (as well as phone companies and magazine publishers) adopted flat-rate pricing. Furthermore, as Appendix C argued, many people believe and expect that information should be free.[4] The last, best hope of the micropayment market is that people will download compressed music files over the wires—to which end the music industry launched its Secure Digital Music Initiative.

The Secure Digital Music Initiative

Buried in ISO's MPEG-2 standard was a once-obscure standard for compression of music, popularly known as MP3—efficient enough to squeeze the average popular song into 5 or 10 megabytes. In the last year, a great deal of popular music has been compressed and distributed over the Web, helped, in large part, by the introduction of portable MP3 players. This has caught the attention of the music

[3]Kevin Werbach (1999) writes that start-ups such as Beenz.com and Flooz.com are trying to enter the micropayment market indirectly. The former is trying to create artificial currencies to consolidate the cyber equivalent of frequent-flyer miles; the latter is focusing on gift certificates. As of late 1999, each was serving roughly 100,000 customers.

[4]Echoed by Ken Casser (1999) of Jupiter Communications in a news article.

industry. The Recording Industry Association of America (RIAA), in particular, has feared the day when only fools would buy the music they could otherwise download. In response, the RIAA prompted the formation of the SDMI (159 members as of January 2000), which generated the SDMI Portable Device Specification in just five months (Version 1.0, 8 July 1999). This specification is a set of requirements (rather than an implementation method). SDMI is a payments standard in the sense that, without it, the RIAA fears that its members will soon see no payments.

The SDMI specification suggests that music will be sold encrypted and watermarked.[5] Such content (1) could be played only on SDMI devices, (2) could be copied only a limited number of times, (3) would be traceable back to the original owner, (4) and would be sold both on line and through stores. It is unclear whether all these are possible if SDMI is to be played on a general-purpose computer (as opposed to a box-and-speakers arrangement, which cannot be tapped into). Software could capture the unencrypted bit stream on the way to the sound card,[6] create a file, compress it, and generate an MP3 file therefrom. If the watermark remains, the original purchaser could be determined from the file and put at legal risk—*if* no hacker figures out how to scrub watermarks away. The difficulty lies in putting a watermark onto a store-bought CD-ROM—more specifically, in persuading music stores[7] to go into the business of imprinting CD-ROMs (with the buyer's name stitched into them), rather than just selling them. Failing that, SDMI music would be sold only on line (as bits or as orders through Web stores), thereby forgoing a CD-ROM market that looks to be larger than the Web market for

---

[5]Why is music the first place that watermarking and other cryptographic techniques are being used to protect intellectual property? Text, and to some extent imagery, has too few bytes to watermark effectively (and most text is sold in physical, and hence analog, form). Video (i.e., DVDs) is already compressed into five-gigabyte files, which currently take too long to download and therefore distribute at Web speed.

[6]SDMI devices that include sound cards would output a waveform to the speakers. One could capture the waveform, run it through an analog-to-digital device, and create a stream of bits—a lot harder then just finding some piece of software to capture the bits flowing among ports. However, marketing SDMI devices that do not work unless they replace sound cards may be a tougher challenge than selling SDMI devices that can use existing sound cards.

[7]Presumably, persuading the fewer and larger mail-order CD-ROM vendors to do this is easier.

years to come. If SDMI, or something like it fails, the market for micropayments would have that much less of a lease on life.

Secure Sockets Layer (SSL)

So, electronic transaction mechanisms using SSL to authenticate merchants and hide transactions from eavesdroppers was left to become the de facto standard for E-commerce. SSL came from Netscape, which put it into browsers and released the protocols to the rest of the world. In essence, the SSL protocol protects HTTP transmissions by adding a layer of encryption (using public-key cryptography licensed from RSA Inc.), thereby hardening it against sniffing by others. Authenticating the server (e.g., is that really Dell Computer's or some trickster's Web site?) also requires a public-key infrastructure—so that the user can go to a trusted party and get the public key by which the complete transmission can be verified as authentic and unaltered. Server authentication has not proceeded very far.

Cryptographic Codes

The easy success of SSL, which, in effect, is a cryptographic standard, belies the raucous history of cryptographic standards to date.

At the risk of oversimplifying a complex and edifying tale, the story suggests the problematic nature of U.S. public policy on encryption. Prior to the mid-1970s, cryptography was essentially the preserve of the National Security Agency (NSA). But in 1977, responding to commercial pressures (e.g., banks that wire money), NIST produced the Data Encryption Standard (DES) for public use.[8] This standard was adopted by industry for a wide range of applications. Fifteen years later, the advent of digital telephony, the burgeoning Internet,

---

[8]In the 1970s, Whitfield Diffie and Martin Hellman, working outside NSA, developed the theory of public-key encryption (PKE). Thus, they solved the problem of passing keys in the clear (which is to say, between two people with no mediating institutions between them) and digitally signing messages. The Diffie-Hellman theory was converted into working code by RSA Inc., whose patented algorithms form the primary commercial counterweight to the NSA. So far, the government has worried less about PKE than symmetric encryption because PKE is computationally difficult to apply to long messages; it is usually used to pass symmetric encryption keys.

and the introduction of inexpensive systems capable of cracking the 56-bit DES created a dilemma within the government. On the one hand, it needed a better encryption standard for its own uses; on the other, it feared that the proliferation of a *public* encryption standard would allow everyone else to keep secrets from the government.[9] The proposed solution, the Clipper chip, used an 80-bit key, part of which contained a Law Enforcement Access Field, whose bits could be revealed to the U.S. government, with proper legal authorization. With these bits known, the rest of the key would be easy for government computers to discover, and the message could be decrypted. Had the proposal succeeded, the government and its vendors (e.g., defense contractors) would have had their own standard. At worst, their exit from the rest of the commercial world would have eliminated a large customer and a natural fulcrum for a standard that would go beyond DES. At best, it would have created a standard that made communications more transparent to the government. But Clipper burned in the subsequent public firestorm.

On January 2, 1997, NIST announced a competition for developing an Advanced Encryption Standard (AES) to replace DES—a sure sign of Clipper's death. (NIST, no date.) The process of selecting and implementing the AES was envisioned as a multiyear activity that would allow extended public comment and a smooth transition away from use of DES. To quote its initial announcement, "AES will specify an unclassified, publicly disclosed encryption algorithm capable of protecting sensitive government information well into the next century." (NIST, no date.) The AES, it went on, was (1) to be publicly defined, (2) to use a symmetric block cipher, (3) to be designed so that its key length can be increased as needed, (4) to be implementable in both hardware and software, (5) and to be available either for free or under terms consistent with ANSI patent policy. Selection criteria included security (i.e., the effort required for cryptanalysis), computational and resource (e.g., memory) efficiency, hardware and software suitability, simplicity, flexibility, and licensing. On August 9, 1999, five semifinalists were chosen to enter into the second round of assessments.

---

[9]An interim solution, the triple use of DES, offered a level of security often considered tantamount to an 80-bit DES, but its development avoided controversy because it exploited an extant standard and lacked a key-recovery mechanism.

The AES was designed to overcome the original limitations of DES, which weakened it against attack and irritated professional non-government cryptographers, who thought the limitations undesirable. It would also establish a new standard for securing some types of E-commerce, and greatly simplify the activities of financial and banking institutions. Finally, its process would test the government's willingness to participate in an open discussion in which core government equities of national security and law enforcement would have to be balanced with privacy concerns. The openness of the process was designed to avoid some of the issues associated with alterations of the original DES algorithms that have raised the suspicions of academic cryptographers over the years. Unlike DES, much of whose testing took place beyond public scrutiny, AES is being run with a far-more-open evaluation process. The new process represents a serious effort to address technical, political, and economic concerns.

AES is a high-stakes game; not only might the standard become a building block for many economic activities, but its process has to satisfy multiple constituencies. Computer scientists (e.g., academic cryptographers) see the AES as their first real chance to deploy widely algorithms developed without inputs from government cryptographers, as well as a process for subjecting competing systems to open large-scale testing. Software vendors see the AES as a chance to expand their markets, dump a now-insecure standard (DES), and come up with a standard acceptable for dealing with the government—and perhaps even overseas customers.[10] Business users see AES as a better way to protect communications and data, while increasing user confidence in the system. Because DES is now used for many financial transactions, the risk that the financial and banking industry may be exposed to penetration of its communications means that it has great interest in developing stronger communication security.

The government generally agrees that AES is needed for both internal uses and to bolster private security. But as a general rule, the government's interests are divided between the need to keep secrets and

[10]Even if AES-based programs cannot be sold overseas, its public definition facilitates building compatible systems.

the need to crack them—security services are interested in protecting communications but at the same time are concerned about preserving their own ability to read arbitrary communication traffic. This duality attracts the greatest attention from political activists who view government action very skeptically. Overall, the activists have had few serious complaints about the AES selection process as such, largely because it is open and well-scrutinized. It helps to have had experience with selecting a similar standard and a memory of the friction that the last attempt entailed.

Cryptographic Procedures

Code crackers will often admit that otherwise unbreakable encryption methods can be hacked by attacking the weak points in the *process* by which plain text is converted into cipher text.

To enhance the cryptographic process, IPsec was proposed in 1995 (RFC 1825) for securing the internetworking layer (IPv4 and IPv6). At that layer, IPsec provides security and authentication for messages in the network. Although it provides no application-layer authentication, it does support an infrastructure to do so. IPsec evolved to RFC 2401, plus a set of supporting RFCs focusing on such issues as encryption. Taken together, the RFCs provide access control, connectionless integrity, data-origin authentication, protection against replays (a form of partial sequence integrity), message confidentiality (encryption), and limited traffic flow confidentiality.

IPsec has attracted limited attention outside of the technical and business communities. As an internetworking issue at heart, it has avoided some of the contentious elements surrounding encryption, which was segregated into a supporting standard. Its controversies have ranged over technical characteristics rather than the clash of rights. Unfortunately, the process has been slow, reflecting the newfound complexity of IP networking and the greater care being taken not to deploy an improperly defined standard on a large scale. The business community has also found IPsec to be largely uncontroversial; network service providers see little reason not to adopt IPsec when it matures. As a general rule, as long as the encryption element is not driving the standards discussion, procedural standards, such as IPsec have been able to ripen unencumbered by serious political fights.

PRIVACY

Privacy is considered by industry as a nice-to-have but not need-to-have feature of E-commerce. If customers demand it, companies will supply it—not necessarily enthusiastically (after all, customer lists have resale value), but willingly enough. But the onus on this side of the Atlantic is on the customer's caring enough about privacy to make it an important factor in patronizing a Web site. Here, standards are meant to enable informed consumer choice. On the other side of the Atlantic, the onus is on those who collect the data, and thus standards are proscriptive: Certain actions are simply forbidden. As it is, the Web does not respect oceans very much, and rules do not pull up as suddenly on the water's edge as legislators might like. And so, the stage is set for a trade row over, yes, standards (or at least standard frameworks).

Platform for Privacy Preferences (P3P)

One sally into the world of privacy standards is the W3C's P3P, a mechanism to automate the control of private information exchanged in an electronic transaction (including site visits). A first phase was completed in late 1997, with a full recommendation expected by the middle of 2000.

The standard's key elements include the disclosure of the site's privacy practices, the expression of the user's preferences, and a medium for negotiation between them. A controlled and secure exchange of data follows an agreement. This is accomplished by using a standard syntax for transmitting the promises made to through the transaction and finding a way to present the data to the two agents for interpretation and negotiation. P3P standardizes eight transfers: requests for data, practice, and preferences; transfer of practices and preferences; requests to transfer data; agreement; and data transfer itself. Negotiations are based on a comprehensive, yet still manageable, range of consumer preferences and on allowing the user's agent to negotiate with the Web site. P3P uses XML and RDF to format the metadata format for such operations; it assumes they are used correctly. While not requiring digital signatures, P3P can use them seamlessly, thereby providing relatively strong authentication of identity for both certificate holders.

P3P exemplifies policies advocated by such groups as the Electronic Frontier Foundation. The point was to have a mechanism for either automatically controlling or allowing the user to control the type and amount of information that is passed. The key in creating this is to define a standardized set of protocol elements. Engineers view P3P as a key step in creating the necessary infrastructure for agent-based transactions extending well beyond privacy concerns.

In a realm where many standards are self-enforcing (i.e., nonstandard products can neither read nor write to the rest of the world), the P3P standard, ultimately, is not. It facilitates the negotiation of information, but not the enforcement of any contract terms.

TRUSTe

In 1996, TRUSTe established itself as a nonprofit organization whose logo could be displayed by Web sites that purchased a license and abided by a privacy code. Such a code would govern what personal information is being gathered; how it is being used; with whom it is shared; and how it will be safeguarded, maintained, and updated. TRUSTe also oversees vendors and supports a dispute resolution mechanism to which consumers can complain about licensed sites. Sanctions under the license range from "forcing a compliance by a CPA [certified public accountant] firm revocation of the trustmark, termination from the TRUSTe program, breach of contract proceedings, or referral to the appropriate federal authority."

TRUSTe is complementary to technical systems, such as P3P, in that it monitors the promises made via the P3P protocol; after all, the technical system has no way to know that a site is lying about how its data are being used. Because TRUSTe's behavior standards have no interesting technical component—it lives in the realm of contact law and tort—it has faced little technical controversy.

Many merchants hope TRUSTe helps them avoid government regulation. As such, TRUSTe has been supported by major Web site operators, which collectively account for a large fraction of all daily visits—but only a small fraction of total Web sites. Activists have had mixed reviews of TRUSTe. The Electronic Frontier Foundation and others have passed on the concerns of activists and have shaped TRUSTe's requirements.

Yet calls for regulation persist. TRUSTe's coverage is limited, and the compromise of information is still an issue. Overall, public concerns over privacy have put pressure to regulate in this area on the Federal Trade Commission and Congress.

EU Directive on Data Protection

The Europeans, as is their wont, have taken a more regulatory approach. The EU Directive on Data Protection (Directive 95/46/EC) requires European entities to provide consumers a variety of protections and effectively limits transactions (except under limited circumstances) with countries lacking similar statutory protections. It also establishes a regulatory framework to control consumer data and standard mechanisms to be used when updating such data. The directive enumerates what types of data require special protection and establishes basic data principles based on fairness, relevance, accuracy, specification of purpose, and retention period. Data must only be collected with informed consent of the provider, except when other considerations, such as contractual obligations, legal requirements, vital interests (e.g., medical information), public interest, and other criteria, overwhelm the specific interests of the data provider.

Several implications followed. First, EU countries will have compatible privacy protection regimes that incorporate certain basic principles. Second, EU countries can no longer restrict their information flow to *other* EU countries based on differing privacy restrictions. Third, countries with noncompliant privacy mechanisms, such as the United States, are theoretically subject to having data flows blocked. In practice, the United States and the EU have approached agreement on safe-harbor principles.

The directive has elicited little response from U.S. technologists; such rules are considered exogenous forces that establish the parameters for their work. Standards, such as P3P, could operate effectively in this regime and may yet do so in Europe. But the directive does *not* make allowances for systems like TRUSTe.[11]

One lesson for standards should be clear. Technical standards for disclosures and agreements are a useful way of mediating claims, in

---

[11]See Blackmer (1998) for a good overview of the directive from a legal perspective.

this case, privacy claims. The EU directive is an existence proof that the political forces pushing an alternative approach, via regulation, do exist. But the ability of a voluntary contract approach to fend off a regulatory approach is directly related to how widely standards are used, a matter of no small importance if contracts are literally executed at click speed. Standards, as such, have to be transparent to the process but clear to the users, and they have to be used widely enough before they can be considered an effective alternative to regulation. The alternative to standards, here, is not less uniformity but, thanks to the threat of regulations, possibly more.

CONCLUSIONS

In the end, standards are a form of soft power; they are effective only insofar as it is in each person's interest to work with the rest. They cannot, themselves, overcome strong contrary inclinations. As long as credit cards are easy for consumers to use, credit-card companies have an uphill battle to ease them to something more complicated albeit less prone to fraud.[12] If consumers do not like being nickel-and-dimed on the Web, micropayment mechanisms will not change that. With most of the student-aged population on the Web, it is difficult to shut off the recirculation of MP3 music even if its possession is almost always a copyright violation. Which is all to say that even the best lubrication cannot make an object slide uphill.

---

[12]Indeed, the coming battle over "electronic wallets" is an attempt to make credit cards and other financial devices even easier to use by making all the necessary data accessible to the browser. In mid-1999, Microsoft, AOL, on-line stores, and the credit-card heavyweights formed a consortium to define E-wallets by defining the Electronic Commerce Modeling Language.

Appendix E

STANDARDS AND THE FUTURE VALUE CHAIN

The need for and development of open standards to facilitate E-commerce and knowledge organization depend on exactly how such realms are structured. Many believe that the Web will redefine the relationships among customers, producers of goods and services, and their intermediaries. The structure of interactions, whether for commerce or knowledge, dictates what information each party needs to exchange and, thus, the terms needed to exchange it—a development that throws into further flux the already difficult task of standardizing on such terms. If nothing else, the self-defined communities busy with their definitions may not necessarily exist as such five and ten years later.

Such possibilities, whose motions can be glimpsed but not their outcomes, lead to the question. What forces in the marketplace have the potential to alter the way future standards are developed?

In the physical world, intermediaries populate the "value chain" to provide key functions or services:

- aggregating buyers and sellers (e.g., wholesalers and retailers)
- reducing the transaction risks among them (e.g., payment methods and guarantees)
- providing information (e.g., advertising and other marketing) on products
- helping consumers select them (e.g., sales agents and clerks)
- customizing them (e.g., changes and alterations)
- forwarding them (e.g., shipping, delivery).

Many predictions have been made about how much "disintermediation" may occur as markets in E-commerce emerge. Intermediaries that depend on information asymmetries to achieve rents, such as mortgage brokers who leverage their knowledge of lender programs, would seem to be at risk as information flows more freely. To survive, they would have to change the medium of their business yet not cannibalize their current market.

Others argue that E-commerce favors a growth in intermediaries, as the affluent would pay a premium for additional customer service. Some new Web businesses provide entirely new classes of service enabled by information technology. Consider Peapod's home delivery of groceries. It takes Web-based orders from customers and sends workers to affiliated chain grocers (such as Safeway) for goods, delivering them to customers at a time prearranged through the Web delivery system. Webvan, by contrast, avoided reliance on grocery store chains and has constructed regional distribution centers (not unlike Amazon.com). Peapod has the advantage of a logistics base whose product mix is tied to proven local tastes. Webvan can tailor the product mix in its warehouses, creating the potential to leverage from low-margin groceries into higher-margin products. Of course, how much various customers like home delivery is unanswered. But the early quest for a best mix of intermediary functions is certain.

To what extent can intermediaries in a sector be consolidated on the Web? New companies are trying to substitute for or eliminate conventional intermediaries, some of whom are trying to create electronic outlets for their services at the risk of eating into their own consumer bases.

Take cars. Today, automakers sell to dealers for resale to end customers. Dealers exist to facilitate product selection (telling buyers about an automobile's features and letting them test-drive the product), arrange financing, and conduct after-sale repair and maintenance. Dealers give automakers customer feedback about features. They also aggregate customer orders and pay for local advertising. E-commerce allows automakers to provide several intermediary functions directly to customers at low cost. Automakers' Web sites already can provide on-line brochures; price estimation; and, soon, interactive virtual demonstrations. *Or,* product and price data and feedback may be provided by third parties (e.g., carpoint.msn.com or www.autobytel.com). But a physical presence is needed for repair and to provide model cars for inspection and test drives.

Even if dealers survive, what will become of their role and market power? Web sites that describe product features, prices, and news may not undermine dealers but, instead, help them; less time is needed to explain the product to well-informed customers. But what dealers do may vary. Luxury brands (e.g., BMW, Daimler-Chrysler, and Lexus) may become more "forward integrated" with their savvy customers by adding personalized services. Thanks to Web-collected customer preference data, dealers may make house (or office) calls cost-effectively because the car they bring is better suited to what a customer might want. Home delivery may follow. Dealers themselves may work out of their homes as maintenance is outsourced to certified facilities.

The conventional wisdom on intermediaries has swung from predictions of their demise (people can buy Italian olive oil directly from the source) to forecasts of great growth (people will frequent vertical portals to discuss and then *from there* purchase olive oil). Now the only forecast is that change—of some sort—is inevitable. For the automobile industry, E-commerce promises not to disintermediate dealers but to transform their role and force the end-to-end consolidation of the value chain with more outsourced functions.

AGENTS AND BOTS

Agents (programs that scan, filter, prioritize, negotiate, or otherwise assist transactions based on their owners' preferences) and bots (programs that search the Web to answer a specific question) may well facilitate negotiations over price, features, and contract fulfillment.

As value chains consolidate, competitors may realize greater opportunity to define the nature of their relationship with the customer. In recent years, agents have been proposed for E-commerce applications with a varied scope and complexity. Some simpler agents are already being used in large-scale E-commerce transactions. But the technology is still young. Patti Maes et al. (1999) have offered the following taxonomy for agents based on their roles in E-commerce; they are grouped according to whether they

- help buyers become aware of some unmet market need and motivate them by using targeted product information
- generate a set of user preferences based on questionnaires or other information and suggest product alternatives based on these preferences
- help differentiate among suppliers based on preferences identified through product brokering activities and on product availability
- negotiate prices and other terms based on customer and supplier parameters
- execute a purchase transaction and delivery of the product based on available payment and delivery methods
- process user service or return requests, as well as product feedback to suppliers.

One example is collaborative technology, first commercialized by Maes's Firefly, Inc. Amazon.com uses it to inform prospective customers (via E-mail, or upon site visit) about new book, music, or video selections that may appeal to them. It uses a preference engine to survey a user's past purchases, create a profile, match this profile to that of other customers, extract a list of similar customers, and recommend what these others have bought. Thus, Amazon.com can simultaneously gather detailed product preference information from their users and generate new sales leads inexpensively. Customers learn about potentially interesting offerings as well. By contrast, software has traditionally learned about individuals only from their past actions. Of note is that collaborative filtering relies heavily on data taken from a large number of users. *If* collaborative filtering yields real value (and doubts persist over whether such technologies scale well) then bigger E-commerce sites may come to understand users better—the reverse of the normal relationship where smaller merchants offset their higher costs with deeper knowledge of each customer. How preferences data are used, bought, and sold clearly has privacy implications. Another firm, Net Perceptions, has a business model that depends on the collection of user data from multiple sites to target customers with suggested products from merchants who buy the agent software.

Agents have been envisioned as electronic butlers assigned tedious and repetitive tasks by their owners. Such agents must accurately understand their owner's preferences and, more importantly, their calculus for making decisions. A number of simpler agent-based technologies do this by having their owners make serial choices. More work—perhaps by extending the collaborative filtering model—is required before agents can rank and deconflict these choices to select an acceptable course of action.

As another example, Jango, an agent technology found in www.excite.com/shopping, lets a user specify the type of item to be bought and some desired features (when metrics of quality can be standardized). Jango can then help buyers search for laptops using such characteristics as manufacturer, model, price range, processor type and speed, hard drive capacity, random-access memory, CD-ROM type, screen size, and modem speed. Suppliers affiliated with www.excite.com are queried to locate the best fit to such preferences. A list appears of product choices accompanied by a column-by-column comparison of different product features. But further details require that customers proceed to the supplier's Web site.

Further examples: Push programs (e.g., from Pointcast Technologies) feed users product information based on a user-supplied demographic and preference profile. Some software companies include, with their product shipments, agents that linger in the operating system user environment, gather periodic feedback from users as they learn the software, and feed these results back to the suppliers via the Web.

Future agents could be both autonomous and portable—capable of negotiating product selections and making purchases automatically (e.g., a new carton of milk every week).[1] A household agent might maintain the household inventory and purchasing system using bar codes to track household supplies (e.g., food, cleaning supplies) as they are used. Periodically, agents could query various stores to find available products, negotiate prices, order them, and arrange for delivery. All this entails knowing a consumer's at-home schedule (e.g., for perishables) and the seller's shipping schedule. Here, too,

[1] See, for instance, the Cross-Industry Working Team (1995).

privacy is a concern—agents necessarily reveal information about their owners.

Just as intermediation struggles would define relationships among the various levels of the "value chain," how agents are used will define the relationship between vendors and consumers. Consumers have an interest in keeping choice open (e.g., portable phone numbers); vendors, in friction—or stickiness (e.g., *not* forwarding E-mail to an address in another vendor's system). Friction results from consumers having invested in the producer. Agents are a yet another battleground. The more that the structure of a consumer's agent adapts to the particular feature structures of one vendor, and the more that a consumer's information structures (e.g., how consumer preferences are expressed, how a consumer's needs are converted into a purchase decision schedule) follow such feature structures, the stickier the relationship. Agents, meant to empower consumers, may, in fact, do the opposite—once vendors determine how. Conversely, the semantics of such features could be encapsulated in standards to which Web sites may adhere.

STANDARDS

Consider an intermediary (which may well exist already) in the business of helping high school seniors select one among thousands of colleges. Before the Web, fat books could only sketch the relevant attributes of competing schools. Today, weight is no constraint, and far more information can be available.

But what information is relevant? By what means are comparisons to be made? How can schools be persuaded to offer comparable evidence to permit such comparisons? On what terms can third parties, such as alumni or townies, participate in the dialog? How can the veracity or relevance of these third parties be evaluated? These are not easy questions, and the first one to get these right is likely to build agents structured along the lines of a solution. And that one may not necessarily open such methods to potential competitors.[2]

---

[2]Not to mention the ethical tension between evaluating vendors' products and taking their advertising money.

Privacy and trust are likely to loom as even larger issues as the factors that promote stickiness become increasingly personal. The more agents replace people in these transactions, the more automated is the auditing of someone's trustworthiness. This requires that transactions be logged consistently even as major merchants would append proprietary extensions atop standard data sets. Conversely, if the agent, determined to protect its owner's privacy, would prefer staying home, some way of expressing the merchant's offerings—or, for collaborative agents, an abstract of their customer base—is required. The latter, of course, is intellectual property, which is difficult to protect once on the road (databases are too granular to watermark easily). Might there be a neutral forum in which the agents of buyers and sellers can meet? If so, what is required is further work on protocols and the semantics of negotiation. The third party has neither the knowing human eye nor the sophisticated (and futuristic) knowledge base to make roughly right guesses when terms and conditions do not match.

Market leaders are rarely friendly to open standards when they dominate and eager to see them when they do not. It is not necessarily in, say, Amazon.com's interest to expose its wares to the peregrination of pricing bots—not when it is so busy building and leveraging its mindshare and the "community" it hopes to foster. Market leaders are also friendly to standards in layers above and below them so as to use the competition among others to increase choices, lower costs, and broaden the market.[3] As one focus group member observed,

> There are various strategies for use in standardization. Those lacking a dominant market position may seek to use standards to enter the market, pursuing either a "leader," "participant," or "fast follower" strategy. The leader strategy requires that the "leader" both deploy technology and then to initiate standards based upon the deployed technology—and then create extensions to the standards which will become the revised standards in the next round of standardization. The participant strategy is to participate in the pro-

[3]This is only true within limits. Rapidly falling prices for personal computers based on sharp declines in the cost of components and assembly have put pressure on the price of Intel's microprocessors as manufacturers tout the advantages of low price over whoever's products are inside.

> cess, deploying the standard in current and future products—but making no effort to lead the process. The fast follower merely wants the standard so that products with the standard's technology can be deployed quickly. All of these models require cooperation with other participants in the market; the first into the market sets the standards and then cooperation is followed by [a cycle of] compete-cooperate-compete.

This leaves the standards process as a forum of discussion and vetting of technologies—a market function with real economic value. That being so, integrity is critical to the standards process, as reflected in the

- inclusion of stakeholders, such as consumers, as well as technologists, vendors, privacy activists, and regulators
- functional transparency that includes not only technical specifications but also rationales and best practices associated with standards
- mutually accessible architects consistent with the principle that a core of individuals interacting with each other work the issues before vetting before the stakeholders *en masse*.

As for the terms that agents use, they must, of course be commonly recognizable—yet one more reason the demand for a consolidated semantics looms large.

Interoperability can be discussed on many levels: physical, communication syntax (e.g., TCP/IP), knowledge syntax (e.g., XML), domain semantics, and process semantics. Process semantics includes the ability to use domain semantics and an adequate representation of each party's interests in dialog. This is the stage that all this intermediation shuffling and the advent agents may yet take us to. In the last five years, the world of information technology has settled fights on communication syntax and appears to have hit on a good solution to the problem of knowledge syntax—and with open standards. As noted, semantics is the next battleground, and beyond it one can glimpse future fights at the service level. And there is no proof that the wars will end with action on that battlefront.

Appendix F

ON THE MEANING OF *STANDARD*

The word *standard* refers both to performance standards (something is at least this good) and—the sense used in this work—conformance standards (this can work with that). It is possible for one thing to be standard in the sense of meeting certain qualifications. But it is impossible for one thing alone to be standard in the sense of working with others. Or put another way, a conformance standard is any convention that is sufficiently common to permit interaction.

The vast world of standards can be divided in many ways:

- What goals do they foster? *Interoperability* standards permit two systems to work together—a necessity, for instance, with Internet devices. *Data exchange* standards help material generated by one system (e.g., a spreadsheet) to be correctly interpreted by another. *Portability* standards ensure that software can run on different platforms; operating systems and computer languages tend to fall in this category. In recent years, such distinctions have begun to erode. The Java language, which is interpreted and run in real time, facilitates portability for software for embedded devices but, when passed over the Web as an applet, promotes interoperability.
- Standards run the gamut from de jure to de facto. A *de jure* standard is a well-documented convention, agreed to by participants in a formal standards forum, such as the ITU. What constitutes "formal" is often a matter of opinion: The ITU is sponsored by the United Nations; the W3C is a large ad hoc membership group, but it, too, has rules. A *de facto* standard is simply a convention in common use.

- Standards also vary from *open* to *proprietary*. Open standards are documented for all, have been developed and are maintained by peers and in public and are supposed to be vendor-neutral. Proprietary standards are developed and maintained by one entity, may require a license to use, and may be incompletely documented.
- Standards may be said to be *light* or *heavy* depending on how detailed their specifications are (particularly when first released). As a correlated characteristic, standards may also be *minimalist* or *structuralist*. Minimalist standards are built up from a set of simple task-oriented primitives (much as words are built from letters). Structuralist standards originate in a reference model of the field of discourse, which is then hierarchically decomposed into functional requirements, each of which is given standard expression.
- Finally, although all the standards discussed here are *voluntary*, others that affect the industry are *mandatory*, especially those that involve over-the-air telecommunications (e.g., the NTSC 6Mz standard for color television).

BIBLIOGRAPHY

Alshuler, Liora, "Xtech '99: Momentum Builds in the IT Sector," at XML.com, March 15, 1999. Available at http://www.xml.com/pub/1999/03/xtech/xtech99.html (last accessed March 23, 2000).

Bearman, David, et al., "A Common Model to Support Interoperable Metadata," *D-Lib Magazine*, January 1999.

Berners-Lee, Tim, *Weaving the Web*, San Francisco: Harper, 1999.

Berners-Lee, Tim, Dan Connolly, and Ralph Swick, "Web Architecture: Describing and Exchanging Data," Cambridge, Mass.: World Wide Web Consortium, June 1999. Available at http://www.w3.org/1999/06/07-WebData (last accessed March 14, 2000).

Blackmer, Scott, The European Union Data Protection Directive, February 1998. Available at http://www.privacyexchange.org/tbdi/EU_PDR/blackmerdirective.html (last accessed March 23, 2000).

Bray, Tim, "XML in XML," XML.com, Web 18, 1999. Available at http://www.xml.com/xml/pub/1999/03/ie5/first-x.html (last accessed March 23, 2000).

Burnard, L., and C. M. Sperberg-McQueen, "TEI Lite: An Introduction to Text Encoding for Interchange," Document No. TEI U 5, June 1995.

Cargill, Carl F., *Information Technology Standardization: Theory, Process, and Organizations*, Rockport, Mass.: Digital Press, 1989.

Casser, Ken, article, 1999. http://www.news.com/News/Item/0.4.33458.00.html (last accessed August 1999).

CEN/ISSS, Electronic Commerce Workshop, European XML/EDI Pilot Project, Project Plan, Draft 1.0, June 5, 1998a.

_____, European XML/EDI Pilot Project, Terms of Reference, June 5, 1998b.

_____, EC Workshop Business Plan V3, May 29, 1998c. Available at http://www.cenorm.be/isss/workshop/ec/documents/Documents98/026.html (last accessed March 23, 2000).

_____, European XML/EDI Pilot Project Mission Statement, October 10, 1998d. Available at www.cenorm.be/isss/workshop/ec/XMLedi/mission.html (last accessed March 23, 2000).

CEN/ISSS, XML/EDI Project Group, "European Comments on the *Preliminary Findings and recommendations on the Representation of X12 Data Elements and Structures in XML* prepared by the X12C Ad Hoc Task Group on the use of XML with X12 EDI, October 1998e.

Chaum, David, "Achieving Electronic Privacy," *Scientific American*, August 1992.

Clinton, William J., and Albert Gore, *Framework for Global E-commerce* , 1997. Available at http://www.iitf.nist.gov/eleccomm/ecomm.htm (last accessed March 14, 2000).

Cover, R., "The Essence and Quintessence of XML: Restrospects and Prospect," December 31, 1998. Available at http://www.oasis-open.org/html/essence_of_xml.html (last accessed March 23, 2000).

Cover, Robin, OASIS, The XML Cover Pages: SML Linking and Addressing Languages (XPath, XPointer, XLink), last update February 21, 2000. Available at http://www.oasis-open.org/cover/xll.html (last accessed March 23, 2000).

Cross-Industry Working Team, *Visions of the NII: Ten Scenarios*, November 1995. Available at http://www.xiwt.org/documents/

Scenarios.html (last accessed the July 12, 1999, version on March 14, 2000.)

Darnton, Robert, "The New Age of the Book," *New York Review of Books*, March 18, 1999, pp. 5–7.

Data Interchange Standards Association, ANSI ASC X12/XML FAQ Sheet, Draft 0.2, July 21, 1998. Available at http://www.disa.org/x12/x12xmlfaq.html (last accessed April 7, 2000).

David, Paul A., and Shane Greenstein, "The Economics of Compatibility Standards: An Introduction to Recent Research," *Economic Innovation and New Technology*, Vol. 1, 1991, pp. 3–41.

de Rose, Steve, interview by Martin Libicki on November 10, 1992.

Dempsey, Lorcan, *Libraries, Networks, and OSI*, London: Meckler, 1992.

Duffy, James, "Vendors cautious about SNMPv3," *Network World*, Vol. 15, No. 9, May 11, 1998, pp. 1, 8.

ECMA—*see* European Computer Manufacturers Association.

EPIFOCAL, EI.pub, "EDI and XML," no date. Available at http://inf2.pira.co.uk/top032.htm (last accessed March 23, 2000).

European Computer Manufacturers Association, "Standard ECMA-262 EXMAScript Language Specification," 3rd ed., December 1999. Available at http://www.ecma.ch/stand/ecma-262.htm (last accessed March 14, 2000).

European Union, Electronic Commerce: A Catalyst for European Competitiveness, Com[99], draft communication, May 10, 1999.

Faculty Taskforce, College of Agriculture and Life Sciences, Division of Biological Sciences, and Albert R. Mann Library, Cornell University, "Journal Price Study of Core Agricultural and Biological Journals" November 1998. Available at http://jan.mannlib.cornell.edu/jps/jps.htm (last accessed March 23, 2000).

Galbraith, I. A., and G. W. Galbraith, "EDML--Electronic Data Markup Language Specification," Version 0.6, January 19, 1998. Available at http://swww.edml.com/edml_061.html (last accessed March 14, 2000).

Garfinkel, Simson, "The Web's Unelected Government," *Technology Review*, November/December 1998, pp. 39-47.

Ginsparg, P., "Winners and Losers in the Global Research Village," presentation at Electronic Publishing in Science Conference, UNESCO, Paris, February 21, 1996. Available at http://xxx.lanl.gov/blurb/pg96unesco.html (last accessed March 14, 2000).

Glenn, Robert, et al., "Wireless Information Technology for the 21st Century," draft white paper, Information Technology Laboratory, NIST, February 17, 1999.

Glushko, R. J., et al., "An XML Framework for Agent-Based E-commerce," *Communications of the ACM*, Vol. 42, No. 3, March 1999, pp. 106–114.

Graphic Communications Association, "What is SML," March 7, 1999. Available at http://www.gca.org/whats_xml/default.htm (last accessed March 14, 2000).

Gruman, Galen, "E-commerce not ready for SET," Computerworld, June 29, 1998. Available at http://www.computerworld.com/home/features.nsf/All/980629qs (last accessed March 14, 2000).

Harbinger Corp., Techniques & Methodologies Working Group, "Frequently Asked Questions," updated November 3, 1999. Available at http://www.harbinger.com/resource/klaus/tmwg/faq.html (last accessed March 14, 2000).

Horny, Karen L., "Automation: The Context and Protocol," in Irene Godden, ed., *Library Technical Services: Operations and Management*, Orlando, Fla.: Academic Press, 1984.

Information Infrastructure and Policy, special issue on interoperability, Vol. 4, No. 4, 1995, pp. 231–366.

Isenberg, David, "Rise of the Stupid Network," June 4, 1997. Available at http://www.phonezone.com/whitepaper/stupidnet.htm (last accessed March 22, 2000).

ISO/IEC JTC1 N4811 and N4833 (Summary of Voting on Document JTC1 N4615, Application from Sun Microsystems, Inc. for Recognition as a Submitter of Publicly Available Specifications for Sun's JavaTM Technologies).

Jellifee, Rick, *The XML & SGML Cookbook: Recipes for Structured Information*, Prentice Hall, 1998.

Johnson, Mark, "XML for the Absolute Beginner: A Guided Tour from HTML to Processing XML with Java," *Java World*, April 1999. Available at http://www.javaworld.com/javaworld/jw-04-1999/jw-04-xml_p.html (last accessed March 14, 2000).

Kernighan, Brian, and Dennis Ritchie, *The C Programming Language*, Englewood Cliffs, N.J.: Prentice-Hall, 1978.

Kramer, Douglas, The Java(tm) Platform: A White Paper, May 1996. Available at http://java.sun.com/docs/white/platform/javaplatformTOC.doc.html (last accessed March 14, 2000).

Krol, Ed, *The Whole Internet*, Sebastopol, Calif.: O'Reilly, 1992.

Lagoze, Carl, "The Warwick Framework: A Container Architecture for Diverse Sets of Metadata," *D-Lib Magazine*, July/August 1996. Available at http://www.dlib.org/dlib/july96/07contents.htm (last accessed March 14, 2000).

Lincoln, Tom, John Spinosa, Sandy Boyer, and Liora Alschuler, "HL7-XML Progress Report," briefing delivered at the HL7 SGML/XML Special Interest Group section of *Healthcare '99.*, 1999.

MacAskill, Skip, "IETF Leader Assesses Health of Internet," interview of Phillip Gross, *Network World*, Vol. 10, No. 16, June 8, 1988, p. 27.

Maes, Pattie, Robert Guttman, and Alexandros Moukas, "Agents that Buy and Sell," *Communications of the ACM*, Vol. 42, No. 3, March 1999, pp. 82–91.

National Institute of Standards and Technology, Advanced Encryption Standard Development Effort, information available at http://csrc.nist.gov/encryption/aes/ (last accessed March 22, 2000).

National Institutes of Health, "E-BIOMED: A Proposal for Electronic Publications in the Biomedical Sciences," draft, May 5, 1999. Available at http://www.nih.gov/about/director/pubmedcentral/ebiomedarch.htm (last accessed April 7, 2000).

National Research Council, National Academy of Sciences, *Realizing the Information Future*, Washington, D.C.: National Academy Press, 1994.

Netscape, "Milestones," 1999. Available at http://www.netscape.com/company/about/backgrounder.html#milestones (last accessed March 14, 2000).

_____, "Browser Plug-ins," 2000. Available at http://www.netscape.com/plugins/index.html (last accessed March 14, 2000).

NIH—*see* National Institutes of Health.

NIST—*see* National Institute of Standards and Technology.

Olsen, G., "Internet Explorer 5 Falls Short on Standards Support, Web Developers Forced to Continue Workarounds," March 1999. Available at http://www.webstandards.org/ie5.txt (last accessed April 7, 2000).

Organisation for Economic Co-operation and Development, "The Economic and Social Impacts of Electronic Commerce: Preliminary Findings and Research Agenda," 1999. Available at http://www.oecd.org/subject/e_commerce/summary.htm (last accessed March 14, 2000).

Paskin, Norman, "DOI: Current Status and Outlook," *D-Lib Magazine*, May 1999.

Pear, Robert, "NIH Plan for Journal on the Web Draws Fire," *New York Times*, March 9, 1999, p. C1.

Raman, Dick, "CEFACT: Proposal for a UN Repository for XML Tags Based on UN EDIFACT," August 31, 1998. Available at http://www.edi-tie.nl/EDIFACT/98crp25.htm (last accessed March 14, 2000).

Raymond, Eric S., "The Cathedral and the Bazaar," *First Monday*, Vol. 3, No. 3, 1998a. Available at http://www.firstmonday.org/

issues/issue3_3/raymond/index.html (last accessed March 14, 2000).

_____, "Homesteading the Noosphere," *First Monday*, 1998b. Available at http://www.firstmonday.org/issues/issue3_10/raymond/index.html (last accessed March 14, 2000).

_____, "Open Source Software: The Halloween Document," 1998c. Available at http://www.opensource.org (last accessed March 14, 2000).

Renfro, Patricia E., "The Speicalized Scholarly Monograph in Crisis, Or, How Can I Get Tenure If You Won't Publish My Book?" *ARL Newsletter*, Association of Research Libraries, No. 195, December 1997, available at http://www.arl.org/newsltr/195/osc-sch.html (last accessed March 23, 2000).

Ricciuti, Mike, "Where the Web Will Be Won," CNET News.Com, 1999. Available at http://news.cnet.com/news/0-1005-201-345512-0.html?tag=st.cn.1 (last accessed March 14, 2000).

Schwartz, Michael, Alan Emtage, Brewster Kahle, and B. Clifford Neumann, *A Comparison of Internet Resource Discovery Approaches*, Boulder: University of Colorado, Technical Report CU-CS-601-92, July 1992.

Seybold Publications and O'Reilly Associates, "The Road to XML: Adapting SGML To The Web," October 2, 1997. Available at http://www.XML.com/pub/w3j/s1.discussion.html (last accessed March 22, 2000).

Shapiro, Karl, and Hal Varian, *Information Rules*, Boston: Harvard Business School Press, 1999.

Spring, Michael B., "Information Technology Standards," in *Annual Review of Information Science and Technology*, Vol. 26, 1991, pp. 1–33.

Stanford Digital Library, The Simple Digital Library Ineroperability Protocol (SDLIP-Core), Stanford, Calif.: Stanford University, no date. Available at http://www-diglib.stanford.edu/~testbed/doc2/SDLIP/ (last accessed April 7, 2000).

Therrien, Lois, "A Double Standard for Cellular Phones," *Business Week*, 3263, April 27, 1992, p. 104.

Van Hentenryck, Karen, "Health Level Seven: A Standard for Health Information Systems," *Medical Computing Today*, June 1998; also http://www.medicalcomputingtoday.com/0ophl7.html.

Veen, Jeffrey, "XML: Metadata for the Rest of Us (Part 1),"interview of Tim Bray, July 7, 1997. Available at http://hotwired.lycos.com/webmonkey/tools/97/27/index0a.html (last accessed April 7, 2000).

W3C—*see* World Wide Web Consortium.

Webber, D., and K.-D. Naujok, "UN/CEFACT/TMWG OO-edi Compatibility With XML/EDI," Version 0.92, July 1998.

Weibel, Stuart, and Juha Hakala, "DC-5: The Helsinki Metadata Workshop," *D-Lib Magazine*, February 1998.

Werbach, Kevin, *Release 1.0*, October 18, 1999.

Wireless Application Protocol Forum, home page. Available at http://www.wapforum.org (last accessed March 14, 2000).

World Wide Web Consortium, XML Pointer Language (XPointer), working draft, March 3, 1998a. Available at http://www.w3.org/TR/1998/WD-xptr-19980303 (last accessed March 22, 2000).[1]

_____, XML Linking Language (XLink), working draft, March 3, 1998b. Available at http://www.w3.org/TR/1998/WD-xlink-19980303(last accessed March 23, 2000.

_____, Extensible Stylesheet Language (XSL), Version 1.0, first working draft, August 18, 1998c. Available at http://www.w3.org/TR/1998/WD-xsl-19980818 (last accessed March 22, 2000).

_____, Extensible Stylesheet Language (XSL), Version 1.0, working draft, December 16, 1998d. Available at http://www.w3.org/TR/1998/WD-xsl-19981216.html (last accessed March 22, 2000).

---

[1]Please note that more recent versions of these working drafts may be available on the W3C cite.

_____, XHTML 1.0: The Extensible HyperText Markup Language. A Reformulation of HTML 4.0 in XML 1.0, working draft, May 5, 1999a. Available at http://www.w3.org/TR/1999/xhtml1-19990505/ (last accessed March 23, 2000).

_____, "HTML 4.01 Specification," December 24, 1999b. Available at http://www.w3.org/TR/REC-html40/ (last accessed March 14, 2000).

_____, "HyperText Markup Language Home Page--Previous Versions of HTML," March 10, 2000. Available at http://www.w3.org/MarkUp/#previous (last accessed March 14, 2000).

Wrightson, Ann M., "XML Study Notes: XML and STEP," School of Computing and Mathematics, University of Huddersfield, 1999. Available at http://helios.hud.ac.uk/staff/scomaw/xmlstudy/xmltop.htm (last accessed March 14, 2000).

XML/EDI Group, XML Repositories Templates Agents EDI "The e-Business framework," no date. Available at http://www.geocities.com/WallStreet/Floor/5815 (last accessed March 14, 2000).